Lecture Notes in Physics

Edited by H. Araki, Kyoto, J. Ehlers, München, K. Hepp, Zürich
R. Kippenhahn, München, H. A. Weidenmüller, Heidelberg
J. Wess, Karlsruhe and J. Zittartz, Köln
Managing Editor: W. Beiglböck

292

E.-H. Schröter M. Schüssler (Eds.)

Solar and Stellar Physics

Proceedings of the 5th European Solar Meeting
Held in Titisee/Schwarzwald, Germany, April 27–30, 1987

Springer-Verlag
Berlin Heidelberg GmbH

Editors

Egon-Horst Schröter
Manfred Schüssler
Kiepenheuer-Institut für Sonnenphysik
Schöneckstraße 6, D-7800 Freiburg, FRG

ISBN 978-3-662-13631-7   ISBN 978-3-540-48053-2 (eBook)
DOI 10.1007/978-3-540-48053-2

This work is subject to copyright. All rights are reserved, whether the whole or part of the material
is concerned, specifically the rights of translation, reprinting, re-use of illustrations, recitation,
broadcasting, reproduction on microfilms or in other ways, and storage in data banks. Duplication
of this publication or parts thereof is only permitted under the provisions of the German Copyright
Law of September 9, 1965, in its version of June 24, 1985, and a copyright fee must always be
paid. Violations fall under the prosecution act of the German Copyright Law.

© Springer-Verlag Berlin Heidelberg 1987
Originally published by Springer-Verlag Berlin Heidelberg New York in 1987
Softcover reprint of the hardcover 1st edition 1987

2153/3140-543210

Preface

Every three years, a European Meeting on Solar Physics is organized by the Solar Physics Section of the Astronomy and Astrophysics Division of the European Physical Society. The board of the Section, i.e.

N.O. Weiss (chairman)	W. Mattig
A. Ambroz	R.W.P. McWhirter
A.O. Benz	E.R. Priest
F. Chiuderi-Drago	R.J. Rutten
J. Christensen-Dalsgaard	B. Schmieder
F.L. Deubner	E.H. Schröter

acting as a Scientific Organizing Committee, decided to hold the fifth meeting in this series in Freiburg/Titisee. The local organization was put into the hands of the Kiepenheuer-Institut für Sonnenphysik at Freiburg.

The general theme 'Solar and Stellar Physics' was chosen in order to focus the meeting on the 'solar-stellar connection'. Space-borne observations in the UV and X-ray spectral regions as well as careful monitoring of the Ca^+ emission of individual stars from the ground have revealed a wealth of information about stellar activity cycles, chromospheres, coronae and winds. On the other hand, it has been recognized that small-scale structures are the key to understanding activity phenomena on the Sun, heating of its outer atmosphere and, probably, the acceleration of the solar wind.

The proximity of the Sun and the multitude of stars open up the possibility of attacking basic problems (e.g. heating of chromospheres and coronae, the nature of flares and eruptive phenomena, the acceleration of winds, cyclic magnetic activity and dynamos, the interaction between magnetic fields and convective flows) as a common effort of solar and stellar astrophysics. The processes can be studied on their natural spatial and temporal scales in the case of the Sun while other stars provide a wide range of physical parameters (rotation, radius, depth of the convection zone, atmospheric structure, evolutionary stage) which can be used to test hypotheses concerning the physical mechanisms.

It was the intention of the organizers to bring together scientists from both communities, the solar and the stellar, to exchange ideas and results. A total of 171 participants from 20 countries made it possible to reach this aim. To cover a wide range of topics, the meeting was divided into four sessions: The conference started with a general introduction, the second session was devoted to lower atmospheres and convection zones, while outer atmospheres and winds were discussed during the third session. Finally, a session on space-borne observation of the Sun and stars concluded the meeting.

In order to stimulate discussion and interaction among the participants, the scheme of the sessions was somewhat different from the usual meeting style. Each session was directed by two chairpersons, one from the solar and one from the stellar astrophysics community. Contributed talks were omitted in favour of extended discussion periods guided by the chairpersons. The discussions focussed on topics raised by the review talks and on the posters related to the session. Unfortunately, the content of the discussions, which were in part lively and fruitful, could not be included in these proceedings.

About 90 poster contributions were shown at the meeting; the abstracts had been refereed beforehand by the respective session chairpersons. A list of posters is given at the end of this volume.

Many persons contributed to the organization and shared a considerable amount of work. It is a pleasure to thank Mrs. G. Abadia and Mrs. Sh. Bloem, the conference secretaries, Mrs. M.v.Uexküll and Mr. A.v.Alvensleben as members of the local organizing committee and V. Anton, S. Immerschitt and H. Münzer for their help before, during and after the meeting. Other members of the Kiepenheuer-Institut and the staff of the 'Kurhaus am Titisee' worked behind the scenes. Thanks are due to all of them.

The session chairpersons, I. Appenzeller, A.O. Benz, A.H. Gabriel, F. Praderie and R.J. Rutten took the burden of refereeing the poster abstracts, preparing and stimulating the discussion periods - and even helped the technician to sort randomly disordered slides. Their effort was crucial to the success of the meeting.

Last but not least, because no meeting can be organized without financial support, we express our sincere thanks to the sponsors of the conference:

The Deutsche Forschungsgemeinschaft

The Ministerium für Wissenschaft und Kunst
des Landes Baden-Württemberg

Freiburg, September 1987   E.H. Schröter   M. Schüssler

CONTENTS

I. GENERAL INTRODUCTION

WHAT CAN THE SUN TELL US ABOUT STELLAR ACTIVITY?

N. O. Weiss
Department of Applied Mathematics and Theoretical Physics
University of Cambridge
Cambridge CB3 9EW
England

Summary

The solar-stellar connection relates high-resolution synoptic solar observations to observations of magnetic activity in stars with different rotation rates and internal structures. Our knowledge of magnetic fields in stellar convection zones is based on detailed observations of field structures in the Sun but recent measurements of magnetic activity in other late-type stars have extended our understanding of the solar dynamo. These observations have stimulated detailed modelling of processes associated with magnetic activity. Modulation of activity cycles in slowly rotating stars can be inferred from terrestrial data extending over the last 10^4 years, while the evolution of the Sun's magnetic field can be inferred from the behaviour of younger stars.

1. Introduction

The aim of this meeting was to bring together two astrophysical communities and to stimulate research by cross-fertilization of ideas. Magnetic activity has been detected in a wide variety of stars with different masses, ages and rotation rates. The resulting information helps us to explain the behaviour of the Sun, the only star on which magnetic features can be studied in great detail and aspects of magnetic activity can be traced backwards for hundreds and even thousands of years. From such measurements we have learnt that the structure of the solar atmosphere is controlled by magnetic fields and this knowledge helps us to interpret observations of other late-type stars. Conversely, the recent surge of research on stellar magnetic activity has led to a growing awareness of the solar-stellar connection, and has stimulated collaboration between solar and stellar physicists.

In this introductory review I shall emphasize interdisciplinarity. My aim will be to summarize the magnetic behaviour of stars for the benefit

of solar physicists and to show how our detailed knowledge of the Sun
can be applied to stellar problems; in addition I shall mention various
solar-terrestrial effects. Naturally I shall try to avoid trespassing
on topics that will be discussed elsewhere in these Proceedings. It is,
however, important to bring out the main theoretical problems that are
raised by the enormous wealth of observations. We understand the basic
mechanisms that cause magnetic activity but there is no convincing
model of the solar dynamo, let alone of the rapidly rotating BY Dra and
RS Can Ven stars that will be discussed by Rodonó. Particular structures
such as small flux tubes and prominences have been thoroughly investigated
but detailed numerical modelling has only recently begun. As yet more
observations at higher resolution are made, from the new generation of
ground telescopes and from space, theoreticians will be kept busily
employed.

2. What stars tell us about the Sun

The magnetic fields of Ap stars seem qualitatively different from
those of late-type stars (Moss 1986). Stars with masses less than
1.5 $M_\odot$ have deep outer convective zones and are often magnetically
active. The degree of activity depends on the angular velocity of the
star (though they are relatively slow rotators). Apparently the combination of turbulent convection and rotation allows a global magnetic
field to be maintained by some kind of dynamo. The resulting activity
can be detected in optical, radio and X-ray emission. 80 years ago Karl
Schwarzschild suggested that chromospheric calcium emission in solar
type stars should be monitored to measure their activity and Hale built
the 60-inch telescope at Mt Wilson for this purpose. But the telescope
was used to study distant galaxies and nothing was done until Olin
Wilson began his survey in 1965. Measurements of Ca^+ emission obtained
since then have provided a fertile source of information on activity in
late type stars (Baliunas & Vaughan 1985).
   Magnetic fields have now been detected in about 50 active stars by
measuring the Zeeman broadening of spectral lines (Marcy 1984; Gray
1984; Saar & Linsky 1987). An active star may show fields with a strength
comparable to that in a sunspot over a substantial fraction of its surface. In addition there are starspots, which do not significantly contribute to these lines but lead to significant changes in luminosity as a
star rotates (Baliunas & Vaughan 1985). Magnetic activity is also associated with flaring, which can be observed in radio or X-ray emission.
Finally, the X-ray observations obtained by the Einstein and Exosat

missions have shown that dwarf stars have hot coronae which, by analogy with the Sun, are magnetically heated and that the degree of activity increases with increasing angular velocity (Pallavicini et al. 1981).

Chromospheric Ca^+ H and K emission provides more detailed information (Baliunas & Vaughan 1985; Soderblom 1985). From measurements of nearby stars it is clear that the degree of activity decreases with increasing age and also depends on spectral type. Since activity occurs in patches the measured emission varies as the star rotates and the period can therefore be determined. Ca^+ emission depends both on the rotation rate and on the structure of the star but it turns out that magnetic activity can be related to a single parameter, the inverse Rossby number $\sigma = \Omega \tau_c$, where Ω is the angular velocity and τ_c the convective timescale at the base of the convective zone (Noyes et al. 1984a). Activity cycles are clearly present in a dozen slow rotators, like the Sun, but rapidly spinning stars show much more complicated time dependence.

From such observations we can assemble a description of the magnetic history of the Sun. During its pre-main sequence evolution a star of solar mass passes through the T Tauri phase, which is characterized by violent flaring and activity, as discussed by Montmerle in these Proceedings. This energy may be released by the destruction of a primeval field rather than by dynamo action in the star (Tayler 1987). As the star approaches the main sequence it contracts and spins up. G stars in the α Perseicluster rotate at up to 50 times the solar rate but the angular velocity of similar stars in the Pleiades (about 3×10^7 years older) has fallen by an order of magnitude (Stauffer et al. 1984, 1985). Apparently stars spin down rapidly immediately after they have arrived on the main sequence but thereafter the rotation rate decays more gradually (Rosner & Weiss 1985). The evolution of the magnetic field then depends on two processes. The first is the dynamo that maintains the field, with a strength that depends on Ω. The second is magnetic braking which removes angular momentum at a rate that depends on the strength B of the magnetic field. Simple parametrizations yield fields that decay with $B \propto t^{-\frac{1}{2}}$ (e.g. Mestel & Spruit 1987). Once Ω is sufficiently small we expect cyclic variations of activity like those found in the Sun, with an amplitude that gradually decreases as the star grows older. Note that this account assumes that the magnetic properties of a star with given mass, composition and angular velocity are uniquely determined.

3. What the Sun tells us about stars

Much of our understanding of the gross properties of late type stars

can be inferred from solar observations. In particular, the frequencies
of solar p-mode oscillations have been used to confirm that standard
stellar models are essentially correct. The base of the convective zone
is at 0.7 R$_\odot$ and the frequencies computed for models are in close agreement with those that have been measured (Christensen-Dalsgaard et al.
1985), though the neutrino problem still remains. The internal rotation
rate can be inferred from rotational splitting of these frequencies
(Claverie et al. 1981; Duvall & Harvey 1984; Duvall et al. 1984, 1986;
Brown 1985; Libbrecht 1986). Apparently the angular velocity at the
equator is almost constant in the convective zone but decreases slightly
in the outer part of the radiative zone. Latitudinal differential
rotation persists to the base of the convective zone but decays within
the radiative zone (cf. Rosner & Weiss 1985). It has been argued that
the enhanced splitting of low degree modes implies that the core rotates
at twice the surface rate though it is hard to see how the consequent
shear could persist if there is any significant magnetic field in the
radiative zone (Mestel & Weiss 1987).

Stellar atmospheres can also be described by reference to the Sun.
Chromospheres and coronae will be discussed elsewhere by Hammer and by
Pallavicini but questions such as what mechanisms heat them can only be
answered by comparing model calculations with detailed solar observations.
Reimers will describe stellar winds but I wonder whether astronomers
would have believed in them if the first space probes had not confirmed
Parker's prediction of the solar wind.

The detailed structure of convection and of magnetic features can
only be detected on the Sun. Title will show the images of photospheric
granulation and of supergranular flow that have been obtained by processing Spacelab 2 results. Magnetic fields have structures that run from
the global scale of coronal holes down to the smallest scales that can
be observed. Specific features like prominences and flares or the
intense magnetic fields in sunspots, pores and intergranular flux tubes
would scarcely have been postulated if they had not been observed.

Finally, we have the record of time-dependent behaviour of the Sun,
based on systematic observations made since the time of Galileo. These
measurements have revealed not only cycles of activity (which we can
now detect in other stars) but also the 22-year magnetic cycle which
distinguishes the Sun's magnetic field from that of a planet like the
Earth. Sunspot observations show both regular and irregular patterns of
behaviour. Activity recurs in cycles and the zones of activity migrate
systematically towards the equator. On the other hand successive cycles
are not strictly periodic and seem to be aperiodically modulated. Moreover, fluctuations in angular velocity correspond to waves migrating

from the pole to the equator over 22 years but recurring with an 11-year period at any latitude (Howard & La Bonte 1980). These observations argue for a dynamo rather than an oscillator (Schüssler 1981; Yoshimura 1981).

4. Modelling solar and stellar behaviour

The wealth of solar observations has stimulated a wide variety of theoretical models, which are subject to the constraint that their predictions can be compared with detailed measurements. Zahn will discuss the hydrodynamics of the convection zone. Here there are three-dimensional numerical experiments in the Boussinesq and anelastic approximations that indicate the global structure of large-scale convection (Gilman 1979; Glatzmaier 1985). These models suggest that there are giant cells, elongated parallel to the rotation axis owing to the Proudman-Taylor constraint imposed by Coriolis forces. As a result, the angular velocity tends to be constant on cylindrical surfaces, decreasing with depth in the convective zone. Experiments conducted in space, with thermal convection simulated by electrostatic forces in a dielectric fluid, have strikingly confirmed this picture (Hart et al. 1986).

Photospheric convection is on much smaller scales and observations provide a picture dominated by the dynamics of exploding granules. Some idea of this fragmentation process can be obtained from Boussinesq results (e.g. Jones & Moore 1979) but realistic simulations have to be compressible. The models pioneered by Nordlund (1985) were first used to describe the solar granulation but have since been developed to cover stars like Procyon and Sirius, as Dravins will explain. This work demonstrates how the solar-stellar convection should be exploited.

Magnetohydrodynamic behaviour has been modelled in great detail (Priest 1982). There are studies of the equilibrium and stability of prominences, of waves and heating mechanisms and of isolated slender flux tubes. The purely hydrodynamic models of convection have been extended to include magnetic fields and treatments of magnetoconvection include both simulations of solar granulation and idealized model calculations (Hurlburt & Weiss 1987; Hughes & Proctor 1987).

The dynamo problem always attracts attention and stimulates some controversy. Aspects of dynamo theory will be discussed by Stix. The basic processes are well understood: differential rotation draws out poloidal field lines to generate toroidal flux and cyclonic eddies provide helicity which yields a reversed poloidal field (Parker 1979). Owing to the difficulty of retaining buoyant magnetic flux in a convecting region it has been suggested that the solar dynamo is located in a magnetic layer in the region of convective overshoot at the base of the convective zone (e.g. Spiegel & Weiss 1980; van Ballegooijen 1982). Self-consistent calculations confirm that the dynamo process works (Gilman 1983; Glatzmaier 1985) though they do not reproduce the detailed structure of the solar cycle. In particular, the dynamo waves travel towards the poles rather than towards the equator. Much of the work has focussed on simpler mean field dynamo models, where the helicity is parametrized as the α-effect and many examples of dynamo waves are illustrated in the literature. I should, however, emphasize that dynamo theory remains in a relatively primitive state. We do not know where the solar dynamo is located and there is no convincing model that reproduces details of the solar cycle. It has been suggested that other processes such as axisymmetric meridional circulations should be invoked (e.g. Wilson 1987) but there is no reason to doubt the efficacy of the dynamo mechanism for generating stellar fields. What is needed are better models of the solar dynamo. Once such models are available they can be applied to other slow rotators with magnetic cycles. Then it may be possible to propose dynamo models for more active stars.

5. What the Earth can tell us about stars

The historical record of solar activity since 1610 shows that the magnetic cycles are irregularly modulated. During the latter half of the seventeenth century (the Maunder minimum) sunspots were rare and there is evidence of earlier grand minima in the record of auroral observations. Variations in the abundances of ^{10}Be and ^{14}C provide a better indicator. Magnetic fields in the solar wind modulate the intensity of galactic cosmic rays which are responsible for producing these unstable isotopes. The 11-year activity cycle has been detected from variations in ^{10}Be abundances in an ice-core from Greenland, where annual layers can be recognized and the Maunder minimum is clearly visible. The ^{14}C abundance anomalies reflect the envelope of the activity cycle. This envelope is consistent with recent sunspot observations and has been carried back over the last 9000 years (Stuiver et al. 1986). Grand minima recur irregularly throughout this period, with a characteristic timescale around 200 yr.

Climatic variations may provide another proxy record. There have been many attempts to establish a correlation between historical indicators of the climate and the 11 or 22 year cycles but none of these is convincing. Yet there are tantalising data sets that may contain fossil records of solar activity. The most striking are the laminated Precambrian varves of the Elatina formation from South Australia. There are about 20 000 "annual" layers deposited during an ice age 6.8×10^8 yr ago and modulated with an 11-12 year period: if these varves reflect the solar cycle then they offer the best record that we have (Williams 1981, 1985; Williams & Sonett 1985). On the other hand, the regular modulation with a 320 "year" period differs qualitatively from the sporadic grand minima in the ^{14}C record, shedding doubt on this interpretation (Weiss 1987).

Astrophysicists tend to expect that the behaviour of a star is uniquely determined by its mass, composition and rotation rate. Oscillatory dynamos are, however, complicated nonlinear systems which need not have unique solutions. The solar cycle is not periodic and appears to be an example of deterministic chaos (Ruzmaikin 1986). The pattern of behaviour during grand minima is quite different from that shown by the Sun over the past 270 years and solar type stars presumably spend about one-quarter of the time in a magnetically inactive state. It is not clear which nonlinear processes are responsible for limiting the growth of the magnetic field (Noyes et al. 1984b). One possibility is that the Lorentz force locally balances the Coriolis force, so changing the pattern of convection (Zel'dovich et al. 1983; Jones & Galloway 1987). There is evidence that the solar surface rotated more slowly during the Maunder minimum (Ribes et al. 1987), suggesting that the distribution of angular momentum was different, perhaps because the rotational constraints on giant cells were partially relaxed. If different possibilities of this type exist we should be cautious before asserting that any star exhibits a unique pattern of magnetic activity, even when averaged over many cycle periods.

6. Conclusion

From what I have said it is clear that theoretical speculations rest on observations. The issues I have mentioned have become topical because improved techniques have led to a much wider range of solar and stellar observations. Kneer will contrast the advantages and disadvantages of observations from ground-based observatories and from space. Under favourable seeing conditions observatories like the Pic du Midi and (we hope) the new stations at Tenerife and La Palma can observe the solar photosphere with extremely high resolution. Yet the images obtained by Title and his colleagues from Spacelab 2 surpass anything that had previously been seen and provide a lesson in what can be learnt by carefully processing high quality results. Over the next decade more missions will be flown in space. Bonnet's summary offers us an exciting prospect, with solar and stellar observations in optical, ultraviolet and X-ray frequencies. The results should keep theoreticians occupied well into the next millennium.

Acknowledgements

On behalf of the Scientific Organizing Committee I would like to thank all those members of the staff of the Kiepenheuer Institute whose hard work helped to make this meeting a success. In particular, we are grateful to E.H.Schröter, M.Schüssler and A. van Alvensleben, and to G.Abadia and Sh.Bloem.

References

Baliunas, S.L. & Vaughan, A.H. 1985, Ann. Rev. Astron. Astrophys. 23, 379.
van Ballegooijen, A. 1982, Astron. Astrophys. 113, 99.
Brown, T.M. 1985, Nature 317, 591.
Christensen-Dalsgaard, J., Gough, D.O. & Toomre, J. 1985, Science 229, 923.
Claverie, A., Isaak, G.R., McLeod, C.P., van der Raay, H.B. & Roca Cortes, T. 1981, Nature 293, 443.
Duvall, T.L., Dziembowski, W.A., Goode, P.R., Gough, D.O., Harvey, J.W. & Leibacher, J.W. 1984, Nature 310, 22.
Duvall, T.L. & Harvey, J.W. 1984, Nature 310, 19.
Duvall, T.L., Harvey, J.W. & Pomerantz, M.A. 1986, Nature 321, 500.
Gilman, P.A. 1979, Astrophys. J. 231, 284.
Gilman, P.A. 1983, Astrophys. J. Suppl. Ser. 53, 243.
Glatzmaier, G.A. 1985, Astrophys. J. 291, 300.
Gray, D.F. 1984, Astrophys. J. 277, 640.
Hart, J.E., Toomre, J., Deane, A.E., Hurlburt, N.E., Glatzmaier, G.A., Fichtl, G.H., Leslie, F., Fowlis, W.W. & Gilman, P.A. 1986, Science 234, 61.
Howard, R. & LaBonte, B.J. 1980, Astrophys. J. 239, L33.
Hughes, D.W. & Proctor, M.R.E. 1987, Ann. Rev. Fluid Mech., in press.
Hurlburt, N.E. & Weiss, N.O. 1987, in The Role of Fine Scale Magnetic Fields on the Structure of the Solar Atmosphere, ed. M.Vazquez, Cambridge University Press.
Jones, C.A. & Galloway, D.J. 1987, preprint.
Jones, C.A. & Moore, D.R. 1979, Geophys. Astrophys. Fluid Dyn. 11, 245.
Libbrecht, K.G. 1986, Nature 319, 753.
Marcy, G.W. 1984, Astrophys. J. 276, 286.
Mestel, L. & Spruit, H.C. 1987, Mon. Not. Roy. Astr. Soc. 226, 57.
Mestel, L. & Weiss, N.O. 1987, Mon. Not. Roy. Astr. Soc. 226, 123.
Moss, D.L. 1986, Phys. Rep. 140, 1.
Nordlund, A. 1985, in Theoretical Problems in High-Resolution Solar Physics, ed. H.U.Schmidt, p. 101, M.P.I. für Astrophysik, Munich.
Noyes, R.W., Hartmann, L.W., Baliunas, S.L., Duncan, D.K. & Vaughan, A.H. 1984a, Astrophys. J. 279, 763.
Noyes, R.W., Weiss, N.O. & Vaughan, A.H. 1984b, Astrophys. J. 287, 769.
Pallavicini, R., Golub, L., Rosner, R., Vaiana, G.S., Ayres, T. & Linsky, J.L. 1981, Astrophys. J. 248, 279.
Parker, E.N. 1979, Cosmical Magnetic Fields, Clarendon Press, Oxford.

Priest, E.R. 1982, Solar Magnetohydrodynamics, Reidel, Dordrecht.
Ribes, E., Ribes, J.C. & Barthalot, R. 1987, Nature 326, 52.
Rosner, R. & Weiss, N.O. 1985, Nature 317, 790.
Ruzmaikin, A.A. 1986, Solar Phys. 100, 125.
Saar, S.H. & Linsky, J.L. 1987, in preparation.
Schüssler, M. 1981 Astron. Astrophys. 94, 755.
Soderblom, D. 1985, Astron. J. 90, 2103.
Spiegel, E.A. & Weiss, N.O. 1980, Nature 287, 616.
Stauffer, J.R., Hartmann, L.W., Burnham, J.N. & Jones, B.F. 1985,
    Astrophys. J. 289, 247.
Stauffer, J.R., Hartmann, L.W., Soderblom, D.R. & Burnham, N. 1984,
    Astrophys. J. 280, 202.
Stuiver, M., Pearson, G.W. & Braziunas, T. 1986, Radio carbon 28, 980.
Tayler, R.J. 1987, Mon. Not. Roy. Astr. Soc., in press.
Weiss, N.O. 1987, in Physical Processes in Comets, Stars and Active
    Galaxies, ed. W.Hillebrandt, E.Meyer-Hofmeister & H.-C.Thomas, p. 46,
    Springer, Berlin.
Williams, G.E. 1981, Nature 291, 624.
Williams, G.E. 1985, Aust. J. Phys. 38, 1027.
Williams, G.E. & Sonett, C.P. 1985, Nature 318, 523.
Wilson, P.R. 1987, preprint.
Yoshimura, H. 1981, Astrophys. J. 247, 1102.
Zel'dovich, Ya.B., Ruzmaikin, A.A. & Sokoloff, D.D. 1983, Magnetic Fields
    in Astrophysics, Gordon & Breach, London.

II. LOWER ATMOSPHERES, CONVECTION ZONES

ON THE ORIGIN OF STELLAR MAGNETISM[*]

Michael Stix

Kiepenheuer-Institut für Sonnenphysik,

Schöneckstr. 6, D 7800 Freiburg

Solar and stellar magnetic fields offer a large variety of interesting aspects: their origin, their variation in space and time, their relationship to the thermodynamic state, their stability etc. I shall not cover all of these aspects in this review, but shall concentrate on two problems which are related to the origin of stelllar magnetism. The first is the concept of mean fields, which plays a key role in the theory. The second is a question which has been discussed recently in particular in the context of the solar dynamo: namely the transition layer at the base of the convection zone as the postulated seat of the dynamo.

Other questions, e.g. flux concentration into narrow tubes, differential rotation, the effect of the magnetic field on solar oscillations etc., will be treated only insofar as they touch the two main topics of this presentation.

Of course, the important subject of magnetix flux tubes, in particular all the observational aspects, will be underrepresented in this review. But I feel justified since a number of recent conferences were exclusively devoted to this theme, cf. the proceedings edited by Schmidt (1985), Deinzer et al. (1986), and Schröter et al. (1987).

1. The Concept of a Mean Field

It is no particular problem to define a mean magnetic field as an average over space or time, or over an ensemble. But it is a great problem to formulate the transport properties of this mean field in a quantitative and correct manner.

[*] Mitteilungen aus dem Kiepenheuer-Institut Nr. 281

The analogy to the kinetic theory of gases provides the easiest, although not entirely correct, access. Gas particles are replaced by parcels of fluid, the mean thermal velocity by the r.m.s. fluid velocity, u, and the mean free path by the correlation length, l, or the mixing length of the turbulent fluid motion. Further, the particles of the disolved substance are replaced by the lines of the magnetic force, and the concentration of those particles by the <u>mean</u> magnetic field , $\bar{B}$. The result is a diffusion process, governed by

$$\dot{\bar{B}} = \eta \Delta \bar{B} \quad , \tag{1}$$

where

$$\eta = \frac{l}{3} ul \quad . \tag{2}$$

Of course, we know that the analogy is misleading because the <u>interactions</u> are so much different. There we merely have <u>collisions</u> between the gas particles and the dissolved particles; here we have communication between fluid parcels by the pressure gradient and other <u>volume forces</u>, and <u>induction</u> of magnetic field by fluid motions.

1.1 First Order Smoothing

In a kinematic theory, where the fluid motions are considered as given, it would be sufficient to describe at least the induction process in a correct manner. The recipe which is commonly used is the well-known procedure of first order smoothing. Let u be the turbulent velocity, and

$$B = \bar{B} + b \tag{3}$$

the magnetic field, divided into its mean and fluctuating parts. The induction equation, considered here for the case of perfect electrical conduction, is

$$\dot{B} = \text{curl } (u \times B) \quad . \tag{4}$$

This equation is also divided into its mean

$$\dot{\bar{B}} = \text{curl } \overline{(u \times B)} \tag{5}$$

and fluctuating part

$$\dot{b} = curl \ (u \times \bar{B}) + curl \ (u \times b - \overline{u \times b}) \quad . \tag{6}$$

First order smoothing consists in the neglect of the second term on the right of (6). Then it is relatively easy to solve (6) for b, to calculate $\overline{u \times b}$, and to substitute the result into (5). The case of stationary and isotropic turbulence, with a correlation time τ, is the simplest; it yields the mean field equation in the form

$$\dot{\bar{B}} = curl \ (\alpha \bar{B} - \beta \ curl \ \bar{B}) \quad , \tag{7}$$

where

$$\alpha = - \frac{1}{3} \tau \ \overline{u. \ curl \ u} \tag{8}$$

and

$$\beta = \frac{1}{3} u^2 \tau \quad . \tag{9}$$

Because the vecolity appears in (8) and (9) only in form of mean second order terms, this approximation has also been called "second order correlation approximation", mostly by M. Steenbeck and collaborators, who developed these concepts (Krause and Rädler, 1980).

Of course it is tempting to justify (2) by (9). And indeed we may imagine that magnetic lines of force are pushed around in a random walk just like dissolved particles (Leighton, 1964). But the identification of β with η requires that $l = u\tau$, which may be true for the solar convection zone (in any case it is assumed to be true in the mixing length theory of stellar convection), but which also is just the condition which makes first order smoothing most questionable! This we shall see in the following section. Additional complications arise because, due the effects of stratification and rotation, stellar convection is far from isotropic turbulence.

1.2 Beyond First Order Smoothing

It has long been recognized that first order smoothing is an incomplete theory. Krause (1967, 1968; see also Krause and Rädler, 1980) has formulated an iteration scheme for a complete solution of the induction equation, and has given a prove for its convergence. Other formulations were given by Knobloch (1977, 1978 a,b) and by Hoyng (1985). Here I shall follow the recent work of Nicklaus (1987), who extended Knobloch's work and actually produced some quantitive results. Knobloch considers the operator

$$L = \text{curl } (\mathbf{u} \times \ ) \tag{10}$$

as a stochastic quantity, and writes the mean field equation in the general form

$$\dot{\overline{\mathbf{B}}} = \overline{K} \ \overline{\mathbf{B}} \ . \tag{11}$$

The coefficient $\overline{K}$ is itself an operator, and is expressed as an expansion in terms of ordered cumulants

$$\overline{K} = \sum_{m=1}^{\infty} \overline{k}_m \ . \tag{12}$$

Each cumulant in turn consists in a sum of terms containing m repeated applications of the operator L. Physically, this formalism describes the dependence of the magnetic field at any instant on the field and the flow in the past.

Let us again turn to the case of stationary and isotropic turbulence. The mean field equation then takes the form

$$\dot{\overline{\mathbf{B}}} = (\eta_1 \ \nabla \times + \ \eta_2 \ \nabla^2 + \ \eta_3 \ \nabla \times \nabla^2 + \ ...) \ \overline{\mathbf{B}} \ . \tag{13}$$

Spatial derivatives of arbitrary order enter into (13), in contrast to (7) which was of second order only. Moreover, the transport coefficients η_i are themselves infinite sums, each with contributions from cumulants k_m of arbitrary order. The first order coefficients (8) and (9) are thus "renormalized".

The question of convergence is an important one. As each application of the operator L essentially lasts one correlation time, τ, and as L should have an upper bound, viz.

$$|L| < u/l \ , \tag{14}$$

we see that

$$k_m \sim S^m \ , \tag{15}$$

where

$$S = u\tau/l \qquad (16)$$

is the Strouhal number. Of course, the velocity u in (14) and (16) need not exactly be the r.m.s. velocity, and l need not exactly be the correlation length, but we may reasonably expect that the order of magnitude is the same. Thus we expect good convergence if S is smaller than some critical value, which should be of order 1. That a small Strouhal number is the condition (for high electrical conductivity) under which first order smoothing is satisfactory was already recognized long ago (e.g. Steenbeck and Krause, 1969).

As a special example of a stochastic velocty field, Nicklaus (1987) considered an ensemble of polarized waves, which was also studied by Drummond et al. (1984) and Drummond and Horgan (1986) in a different context (cf. the following section). Each realization of this ensemble is of the form

$$\mathbf{u} = A \sum_{l=1}^{N} \{ [\mathbf{b}_1 \times \mathbf{k}_1 \cos\psi - (\mathbf{c}_1 \times \hat{\mathbf{k}}_1) \times \mathbf{k}_1 \sin\psi] \cos(\mathbf{k}_1.\mathbf{x} - \omega_1 t)$$

$$(17)$$

$$+ [\mathbf{c}_1 \times \mathbf{k}_1 \cos\psi - (\mathbf{b}_1 \times \hat{\mathbf{k}}_1) \times \mathbf{k}_1 \sin\psi] \sin(\mathbf{k}_1.\mathbf{x} - \omega_1 t) \} \quad .$$

Here A is a constant of normalization, chosen so that

$$NA^2 \to u^2 \text{ (a constant)} \qquad (18)$$

as $N \to \infty$; the vectors $\mathbf{b}_1$ and $\mathbf{c}_1$ are random variables distributed uniformly over the unit sphere; the wave vectors $\mathbf{k}_1$ are also isotropically distributed, but with a spectrum $E(k_1)$ where $k_1 = |\mathbf{k}_1|$; $\hat{\mathbf{k}}_1$ is the unit vector $\mathbf{k}_1/k_1$, and ω_1 are random frequencies with a distribution $D(\omega_1)$.

Expression (17) constitutes an incompressible flow. Each term is a _transverse_ wave, and can therefore be polarized. The free parameter ψ defines the degree of circular polarization, or _helicity_.

Let us restrict our attention to the first corrections to first order smoothing which arise in the formalism of cumulants. It can be shown that for $N \to \infty$ (17) tends to a gaussian turbulence where third order correlations vanish, and forth order correlations decay into products of second order correlations. The terms with third and forth order derivatives in (13) also disappear. We formally recover the mean field equation (7), but α and β are renormalized:

$$\alpha = \alpha_2 + \alpha_4 \quad ,$$

$$\tag{19}$$

$$\beta = \beta_2 + \beta_4 \quad .$$

The various contributions to α and β are shown in Fig. 1 for the following wave number and frequency spectra:

$$E(k) = \frac{2\lambda}{\pi} \exp\left(-\frac{\lambda^2 k^2}{\pi}\right) \quad , \tag{20}$$

$$D(\omega) = \frac{1}{\pi\tau} \frac{1}{(1/\tau)^2 + \omega^2} \quad , \tag{21}$$

where λ and τ are the correlation length and time, respectively. Nicklaus (1987) has studied a number of further examples. The result shown here is quite characteristic: α_4 and β_4 may become of comparable magnitude to α_2 and β_2 when $S \to 1$; the forth order terms may be positive or negative, they may even change the sign of α or β altogether. The α terms are proportional to $\sin 2\psi$ and therefore vanish for $\psi = 0$. Of the β terms the second order contribution β_2 in independent of ψ, but β_4 strongly depends on ψ (maximum helicity, i.e. $\psi = \pi/4$, is chosen in Fig. 1 unless indicated otherwise).

The conclusion of these experiments is clear: in order to be reliable the determination of transport coefficients for the mean magnetic field must go beyond first order smoothing. The problem is of course that as soon as α_4 and β_4 become significant contributions, we have also a good indication that, because of (15), higher contributions become significant as well! Thus the situation is difficult indeed, in particular as we have not yet considered the (certainly important) deviations from isotropy and, in addition, have sofar neglected the _dynamic_ response of the velocity field u to the magnetic force.

1.3 Lagrangian Approach

Yet another possibility to avoid first order smoothing has been explored in recent years. First persued by Moffat (1974) and Kraichnan (1976), it takes advantage of the fact that, in Lagrangian co-ordinates, the induction equation has an exact solution (see also Moffatt, 1978). This solution requires the knowledge of the path x(a,t) of a fluid parcel initially at position a, which, in the kinematic case, can be calculated from the given velocity field:

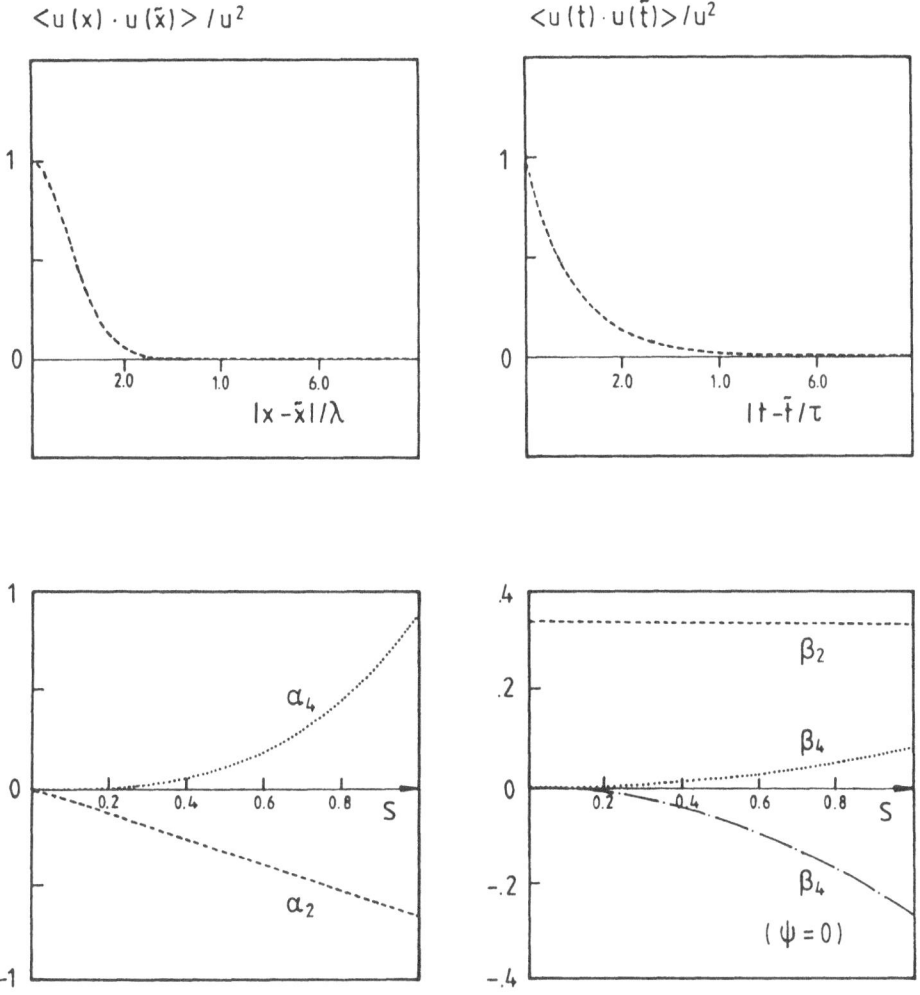

Fig. 1. Correlation in space and time (upper panels) for the velocity field (17),
with spectra (20) and (21), and contributions to the α-effct and turbulent
diffusivity (lower panels), as functions of the Strouhal number, S = uτ/λ. From
Nicklaus (1987).

$$x(a,t) = a + \int_0^t u^L(a,\tau) \, d\tau \quad , \tag{22}$$

where

$$u^L(a,\tau) = u(x(a,\tau),\tau) \tag{23}$$

is the velocity along the path. For an initially homogeneous field, B_o, the k-th component of the Lagrangian solution is then

$$B_k(x,t) = B_{ol} \, \partial x_k / \partial a_l \quad . \tag{24}$$

Having B we may calculate $\overline{u \times B}$ which (assuming $\bar{u}=0$) is equal to $\overline{u \times b}$, the desired term of our mean field equation (5). We find for the i-th component

$$\overline{(u \times b)}_i = \varepsilon_{ijk} \, \overline{u_j \, \partial x_k / \partial a_l} \, B_{ol} \equiv \alpha_{il} \, B_{ol} \quad . \tag{25}$$

This relation defines a tensorial α coefficient. A similar formula, involving an initial field with a homogeneous gradient, defines the corresponding β tensor (Moffatt, 1978). For isotropic turbulence, the case which has been treated sofar, these tensors are reduced to scalar coefficients, namely the α and β which appear in (7).

Of course the coefficients thus defined depend on time, but for a sufficiently long period of integration the magnetic field should forget its initial configuration, and we may hope that α and β become independent of t. The calculations of Drummond and Horgan (1986) indicate that this is indeed the case. These authors used the velocity field (17), with a Gaussian distribution of variance w_o^2 for the frequencies w_n, with $\psi=\pi/4$, and with two forms of the wave number spectrum: a uniform distribution over a sphere of radius k_o (the "δ-shell"), and a Gaussian distribution, of variance k_o^2, for each component of k_l. Drummond and Horgan even included the effect of finite electrical conductivity, i.e. non-zero diffusivity (κ in their notation) into their calculation. Figure 2 shows, as functions of κ, the values toward which the α and β coefficients converge after a long enough integration and after averaging over up to 4.5×10^5 calculated fluid pathes. These results are in arbitrary units; e.g. if $k_o = 6 \text{ m}^{-1}$, $w_o = 10 \text{ s}^{-1}$ and $u_o^2 \equiv u^2 = 3 \text{ m}^2/\text{s}^2$, then α is in m/s, while β and κ are in m^2/s. The parameter values were deliberately chosen to satisfy the relation $w_o \approx u_o k_o$, which corresponds to a Strouhal number of order 1, the interesting case as far as stellar convection is concerned.

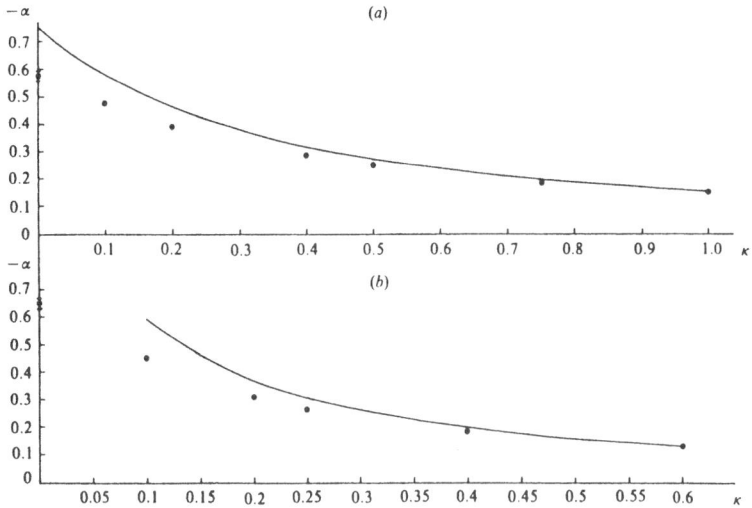

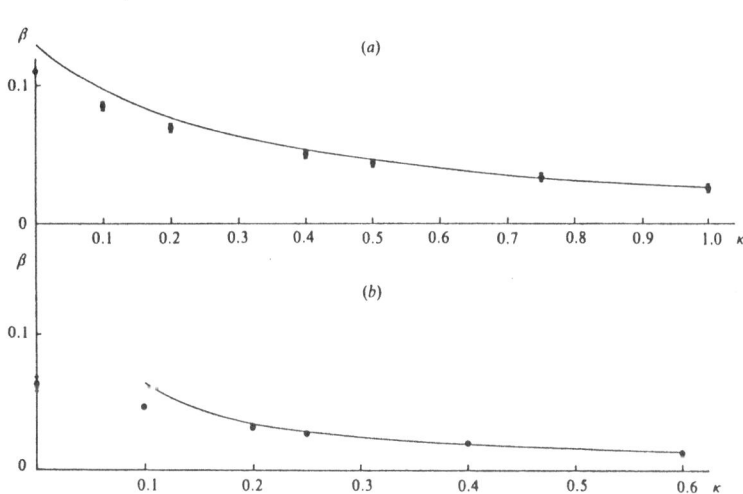

Fig. 2. Coefficients α and β, for t → ∞, as functions of diffusivity, for the case $k_o = 6$, $ω_o = 10$, $u_o^2 = 3$. (a) δ-shell; (b) Gaussian spectrum. The continuous curves result from first order smoothing. From Drummond and Horgan (1986).

We see that even for perfect conductivity ($\kappa=0$) convergent results are obtained. It is particularly encouraging that the results of first order smoothing are not too far away from the exact solution, although the difference is most markedly in the interesting limit $\kappa=0$.

2. The Transition Layer Dynamo

Let me now turn to a quite different problem, namely the question of a hydromagnetic dynamo situated in a transition layer between the outer convection zone and the radiative interior of a star. This question has been discussed recently primarily in the solar context, but of course it is relevant for all main sequence stars which have an outer convection zone.

2.1 Why not in the Convection Zone?

The first answer to this question is magnetic buoyancy and the general instability of fields in the convection zone. Magnetic buoyancy arises because in a magnetized plasma the gas pressure is partly replaced by magnetic pressure. In general, therefore, the density is lower than in the unmagnetic environment (Parker, 1955a). This effect, further investigated by Parker (1975), leads to rapid rise of flux tubes embedded in the convection zone. Calculations of Unno and Ribes (1976) and Schüssler (1977) showed that turbulent viscosity opposes the rapid rise, but the common convective instability, due to the superadiabatic mean temperature gradient, accelerates it further. Spruit and van Ballegooijen (1982) found that, in any case, horizontal flux tubes are unstable if

$$\beta\gamma(\nabla-\nabla_a) > -1 \quad , \tag{26}$$

where

$$\beta = \frac{2\mu P}{B^2} \quad , \tag{27}$$

γ is the ratio of specific heats, and ∇ and ∇_a are the actual and adiabatic temperature gradients ($d\ln T/d\ln P$). In a convection zone, the instability occurs regardless how small the field B is. Moreno-Insertis (1983), who included the effect of "convective buoyancy" into his models of rising flux tubes, obtained even shorter rise times than Parker.

The argument here is not that a dynamo would be entirely impossible under these circumstances. However, as soon as some magnetic flux is generated by shearing motions, it would rise to the surface. We would then see a large amount of magnetic flux in form of weak flux tubes distributed over the solar surface, but not the large flux fragments of 10^{13} Wb or more which are typical for the bipolar regions of the Sun (e.g. Garcia de la Rosa, 1987).

The fact that these bipolar regions strictly follow Hale's polarity rules, with practically no exception, may also be interpreted as evidence for a dynamo layer below the convection zone. There we may expect (perhaps overshooting) motions which are less turbulent than in the proper convection zone. Hence the magnetic field may behave more orderly.

Another important argument emerges from hydrodynamic models of convection of a compressible fluid in a rotating spherical shell (Glatzmaier, 1984, 1985a; Gilman and Miller, 1986; for a recent review see Glatzmaier, 1987). The differential rotation obtained in such models has an angular velocity, ω, which is constant on cylinders parallel to the axis of rotation, a consequence of the rotational constraint. With the observed latitudinal gradient of ω, this leads to $\partial\omega/\partial r > 0$. On the other hand, the Coriolis force renders the convection helical with negative helicity in the northern, and positive helicity in the southern hemisphere (except for the lowest part of the shell, where the sign is opposite). The α-coefficient for the regeneration of the mean poloidal field is then positive in the north and negative in the south. In this case the kinematic αω-dynamo yields a mean field migrating toward the poles rather than toward the equator. The dynamic calculations of Gilman and Miller (1981), Gilman (1983), and Glatzmaier (1985a) confirm this result.

A related difficulty is that the dynamic calculations yield an α effect and a differential rotation of comparable magnitude, i.e.

$$|\alpha| \approx |r^2 \nabla\omega| \quad . \tag{28}$$

According to the αω-dynamo, we would then expect mean poloidal and toroidal field components which are also of comparable magnitude, unlike the solar mean poloidal field of order 1 G (measured) which is much weaker than the mean toroidal field of order 100 G (inferred from the flux measured in bipolar regions).

We must now see whether a boundary layer at the base of the convection zone helps to avoid the diffculties mentioned in this section. Spiegel and Weiss (1980) first discussed such a layer.

2.2 The Overshoot Layer

Thermal convection is driven by a superadiabatic temperature gradient, but it may overshoot into a layer of subadiabatic stratification. For the present purpose this is attractive because (26) would allow (although not ensure) stable flux tubes in such a layer.

In the Sun, the base of the convection zone is where $\nabla = \nabla_a$, at a depth of about $0.7\ r_\odot$. At this depth the temperature gradient has a slight variation, which is visible both in a solar model calculation, and also in the result obtained by Christensen-Dalsgaard et al. (1985) who inverted observed p mode eigenfrequencies, cf. Fig. 3, upper panel.

In order to describe convective overshoot we must not use the local mixing length formalism because then the convection velocity would vanish at $\nabla = \nabla_a$. A number of formalisms have been developed to include convection velocities below this level. Van Ballegooijen (1982) considered linear modes of overshooting convection, and Schmitt et al. (1984) treated the overshoot in form of plumes.

Here I adopt the simpler non-local model of Shaviv and Salpeter (1973), which was originally designed to model overshoot in convective stellar cores, but was recently applied to the solar problem by Pidatella and Stix (1986). In this model convective "bubbles" continue to be (negatively, i.e. downwards) buoyant as they cross the level $\nabla = \nabla_a$ because of their acquired (negative) temperature excess, δT (Fig. 3, lower panel; notice that the level where $\nabla = \nabla_a$ is still on the right of the depth range shown). Only at some deeper level δT becomes positive, and the bubble will be breaked instead of driven. From that level on, the convective energy transport is downward, so that the radiative transport must exceed the total. In a solar envelope program such as the one used by Pidatella and Stix the ratio, l/H, of mixing length to scale height is a free parameter. But the value 1.38 used for the present result is taken from a full solar model calculation. In fact, preliminary results of D. Skaley (1987, private communication) of non-local mixing length theory in a full solar model essentially confirm the result shown in Fig. 3. At the base of the overshoot layer there is a sudden transition to the radiative, stably stratified, core.

Van Ballegooijen (1982), and Pidatella and Stix (1986) estimated the magnetic flux which can be stored in form of flux tubes in their respective overshoot layer models

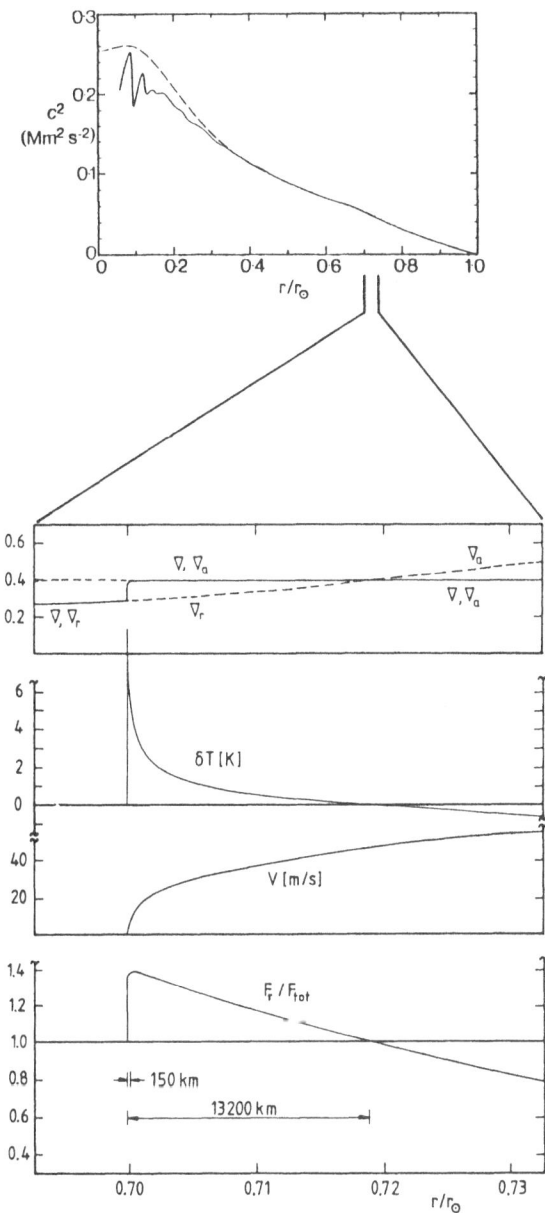

Fig. 3. Upper part: square of the sound speed in the Sun; solar model (dashed) and p mode inversion (solid). From Christensen-Dalsgaard et al. (1985). Lower part: temperature gradients, temperature excess of convective "bubbles", convection velocity, and the ratio of radiative and total energy flux at the base of a model convection zone. After Pidatella and Stix (1986).

and concluded that it is of order 10^{15} Wb, quite sufficient to account for the flux
seen at the solar surface in form of bipolar regions at the maximum of the solar
cycle.

2.3 Where is the Shear?

The generation of a toroidal magnetic field from a poloidal parent field by
non-uniform rotation is a reliable ingredient to all dynamo models of the solar
cycle. If we have an overshoot layer available for the storage of toroidal magnetic
flux, then of course it would be best to have the gradient of angular velocity in
that same layer. One of the first $\alpha\omega$-dynamo models for the Sun, proposed by
Steenbeck and Krause (1969), works with a concentrated shear layer at the bottom of
the convection zone. It has been estimated that a total change of angular velocity
across the layer which is comparable to the total variation in latitude (as seen at
the surface) would be sufficient.

Unfortunately, even such a mild concentration of radial shear so far is not evident
from the rotational splitting of solar p mode oscillations. The equatorial rate of
rotation seems to decrease toward the interior (Duvall and Harvey, 1984), but this
decrease is distributed over a wide range of depth, cf. Fig. 4, upper panel. A
tentative extension of this ω-profile to higher latitudes, also based on p mode
splitting, was presented by Gough (1987). It is also shown in Fig. 4, and disagrees
somewhat from what we would expect from the hydrodynamical models mentioned above:
the cylindrical surfaces of constant angular velocity which, partly below the
bottom of the convection zone, are connected and mark the desired radial shear
(Fig. 4, bottom). DeLuca (1986), whose dynamo model I shall discuss below,
essentially incorporates this picture, which is similar to the one arrived at by
Rosner and Weiss (1985).

It seems that we must wait for better p mode frequencies before final conclusions
concerning the surfaces of constant angular velocity can be drawn. But even with a
wider distribution of shear it should be possible to save the model of a magnetized
layer. This is because there are a number of processes which counteract the above
mentioned tendency of magnetic buoyancy and instability, and lead to a downward
transport of flux: topological pumping, meridional circulation, and a number of
others, as enumerated e.g. by Schüssler (1983). For a new discussion of the
"diamagnetic" effect of a non-homogeneous turbulence see Krivodubskii (1984a).

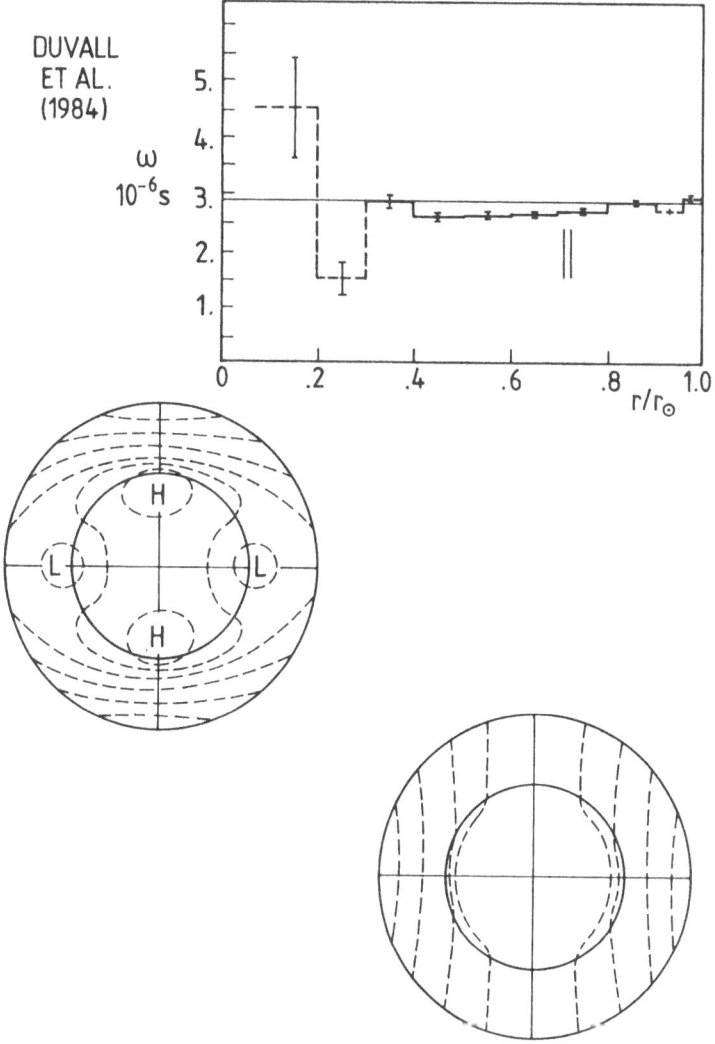

Fig. 4. Top: equatorial angular velocity, as a function of depth in the Sun. The two parallel bars mark the overshoot layer. Middle: tentative shape of constant ω surfaces, from p mode inversion. Bottom: qualitative distribution of constant ω surfaces, based on hydrodynamical models.

2.4 Evidence for a Magnetic Layer?

Spiegel and Weiss (1980) already pointed out that the presence of a magnetic layer at the base of the Sun´s convection zone could alter the convective energy transport, and suggested that luminosity variations of order 0.001 $L_\bullet$ over the solar cycle would be a possible consequence.

Here I like to mention another possibility to see the cyclic variation of magnetic flux inside the Sun: The variation of the p mode oscillation frequencies.

For the period 1980 to 1984, Woodard and Noyes (1985) found a mean decrease of 0.4 µHz for the frequencies of degree 0 to 3, and this has been confirmed by Fossat et al. (1987). However, the effect is of the same order as the accuracy by which the frequencies are known, and Pallé et al. (1986), using a larger number of modes, did not find a significant systematic mean variation over the period 1977 through 1984. Isaak et al. (1987) demonstrated that the l=0 frequencies, on the average, _increased_ rather than decreased, while either case occured among the l=1 frequencies.

It seems hard to believe that the observed frequency variations, if significant, are a direct consequence of a cyclic mean toroidal field in (or below) the convection zone. This is because both Roberts and Campbell (1986) and Vorontsov (1987) showed that a field strength of order 10^6 G is required to produce an effect of the observed magnitude. This is far stronger than the field of order 10^4 G which we expect to be stable in an overshoot layer, according to (26) and (27). Perhaps a small cyclic change of the solar radius plays a role, as conjectured by Woodard and Noyes (1985).

For a _concentrated_ layer, e.g. at the base of the convection zone, there is a more subtle effect: in addition to a mean term, a frequency shift which (for any given l) is _periodic in the frequency itself_. This result has been obtained by Vorontsov (1987) by means of asymptotic theory. The effect is of the same order as the mean shift, and therefore not detectable (for fields of order 10^4 G) with the present accuracy of eigenfrequencies. Nevertheless it is a very interesting perspective, because, as Vorontsov suggests, "with a large number of experimental frequencies availble the periodic signal could be detected well under the noise level".

The effect of magnetic fields on solar p modes of _high_ degree l has been investi-gated by Bogdan and Zweibel (1985) and Zweibel and Bogdan (1986). These modes have their turning points much closer to the surface of the Sun: for a magnetic layer at the base of the convection zone they are therefore of lesser interest.

2.5 The Models of DeLuca and Glatzmaier

Let me now turn to dynamo models actually computed for the layer at the base of the convection zone. The model of DeLuca (1986; see also DeLuca and Gilman, 1986) is kinematic and uses the results of hydrodynamic calculations as prescribed ingredients. The first of these ingredients is a shear layer at the base of the convection zone, with $\partial\omega/\partial r > 0$, as illustrated in Fig. 4 (bottom). The second is an α-effect with $\alpha_{north} < 0$ and $\alpha_{south} > 0$. This type of α distribution is obtained in the lower part of a spherical convection zone, a result which, prior to the dynamical calculations already mentioned, was obtained by Yoshimura (1972). A similar distribution of α is found in a turbulent velocity field with a downwards decreasing turbulent intensity (Krause, 1967; Krivodubskii, 1984b); such is certainly true in the lower part of the convection zone, including the overshoot layer, cf. Fig. 3.

With these ingredients DeLuca's model is almost a classical $\alpha\omega$-dynamo, as first described by Parker (1955b). The product of α-effect and shear is such that the field migrates toward the equator, and that the field of dipolar parity is easier excited than the field of quadrupolar parity. The only difference to the older models is that both α and $\partial\omega/\partial r$ have their signs reversed; the implications of this will be discussed in Sect. 2.7 below.

The model of Glatzmaier (1985b) resembles that of DeLuca, although it is not a kinematic, but a fully consistent dynamical calculation. Unfortunately, a complete magnetic cycle could not be simulated. Nevertheless, as far as the calculation goes it confirms the above-mentioned ingredients, even under the conditions of the non-linear feedback by the magnetic force.

2.6 Schmitt's Model

Schmitt (1984, 1985) derives his dynamo from an instability of a toroidal field in the transition layer with a gradient in the vertical direction, such that $d(B/\varrho)/dz < 0$. The instability takes the form of (slow) magnetostrophic waves, driven by magnetic buoyancy. There is an approximate equilibrium in these waves between the Lorentz and Coriolis forces.

The perturbations **u** and **b** forming the magnetostrophic wave are used to calculate

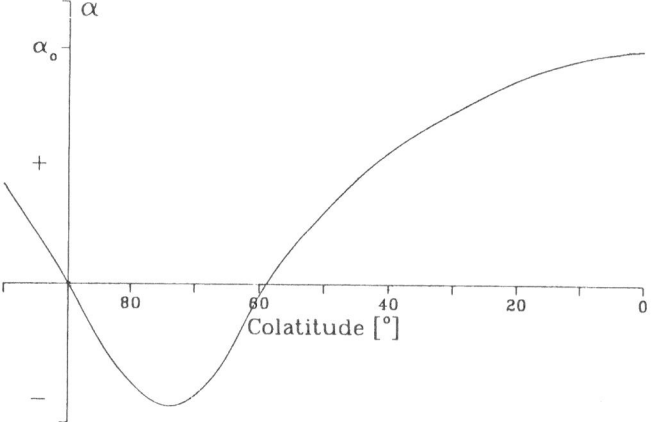

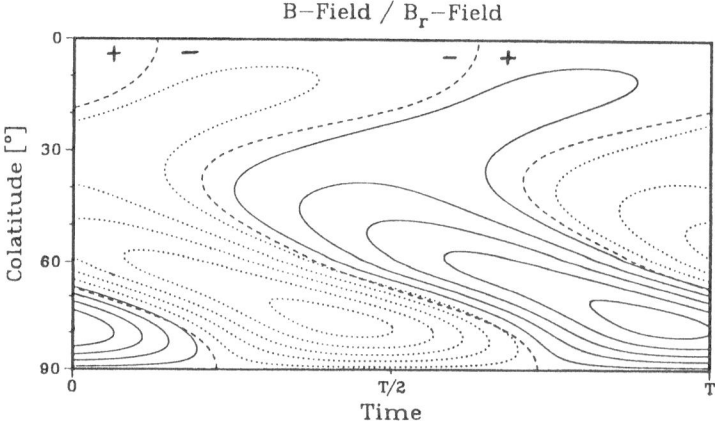

Fig. 5. Upper part: α-effect in the northern hemisphere, as a function of
colatitutde. Lower part: contours of constant toroidal field (solid positive and
dotted negative), and curves of zero radial field (dashed) in a time-colatitude
(butterfly-) diagram. Adapted from Schmitt (1987).

the mean electric field, $\overline{\mathbf{u} \times \mathbf{b}}$, and the ratio of the toriodal component of this electric field to B defines an α coefficient. The α thus found changes sign in latitude, as shown in Fig. 5, upper part, for the northern hemisphere. In the low-latitude zone, where we observe the bipolar groups and therefore infer the strong toroidal field, it has the same sign as the α in DeLuca's and Glatzmaier's models. Accordingly, in order to obtain the equatorward migration in this low-latitude zone, Schmitt (1987) employs an angular velocity with ∂ω/∂r > 0 in his αω-dynamo. The resulting mean field is shown in Fig. 5, lower part. In addition to the low-latitude branch this field shows a second branch, at higher latitudes, which migrates toward the poles; of course, this is a consequence of the change of sign of the α coefficient wihtin each hemisphere.

Partially, Schmitt's dynamo is a _dynamic_ model, because its α-effect is based on an instability of a toroidal field of finite amplitude. It would be nice to close the circle and to identify this field with the toroidal field actually generated in the dynamo. For a stationary mean field (with tentative application to the Earth) such a self-consistent dynamo has been presented by Fearn and Proctor (1984).

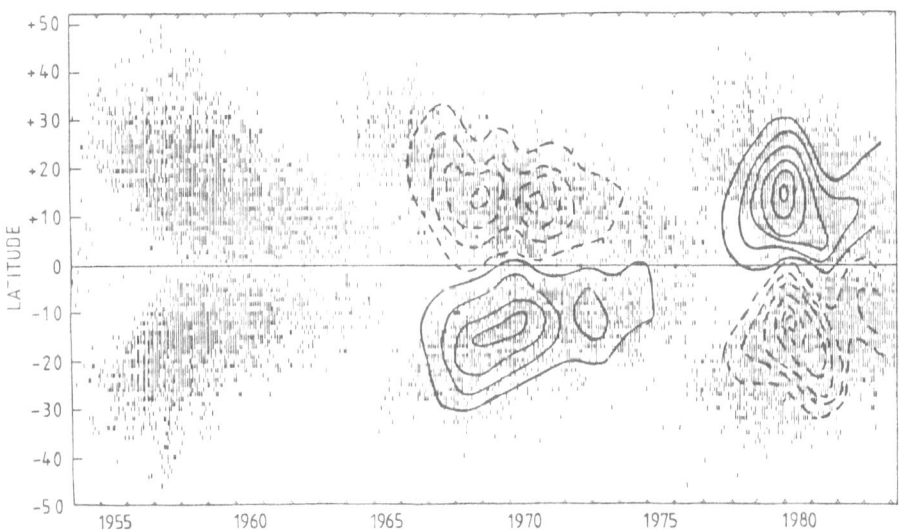

Fig. 6. Butterfly diagram, from Mt. Wilson observatory (courtesy R. Howard), and contours of the mean poloidal field, as obseved at Mt. Wilson and the Wilcox Solar Observatory at Stanford.

2.7 The Phase of the Poloidal Field.

I have pointed out previously (Stix, 1976) that for an $\alpha\omega$-dynamo the signs of both α and $\partial\omega/\partial r$ can be inferred from the observed phase relation between the poloidal and toroidal mean fields. The result was that, in the dynamo region, $\alpha_{north} > 0$, $\alpha_{south} < 0$, and $\partial\omega/\partial r < 0$. Figure 6 extends the range of observed fields until 1984, and confirms the earlier conclusion. This conclusion is in clear contrast to the hydrodynamic results mentioned above, and to the dynamo models described in the two preceding sections.

Thus, the dynamo in the overshoot layer solves a number of problems, but unfortunately creates a new one. It may be stable against fast loss of magnetic flux, it is consistend with hydrodynamics, it does yield the right direction of field migration. And it may even satisfy the condition $|B_{pol}| << |B_{tor}|$ because, as van Ballegooijen (1986) suggests, the efficiency of the α-effect may be reduced due to the fact that the magnetic field is located in a stable layer. The new problem is the wrong phase of the mean poloidal field. Are the observed mean fields not significant? Or is the theoretical model wrong? Yoshimura (1976) suggests that the observed phase relation rather hints toward a dynamo with <u>latitudinal</u> shear, $\partial\omega/\partial\theta$. But, at the same time, his dynamo is located in the upper part of the convection zone, and thus faces the difficulties caused by buoyancy and instability.

3. Other Topics

In this lecture I have concentrated on two points: the mean field concept, and the dynamo in the overshoot layer. I shall briefly review a few more questions in this section, but first refer to the reviews of Schüssler (1983), Stix (1984), Belvedere (1985), Moss (1986), and Weiss (1986, 1987). Most of these reviews also cover the problem of <u>stellar dynamos</u>, and in particular discuss the important relation between the stellar rotation period, P_{rot}, and the period, P_{cyc}, of activity cycles. Noyes et al. (1984) find

$$P_{cyc} \sim (P_{rot}/\tau_c)^{1.25} \quad , \qquad (29)$$

where τ_c is the convective turnover time near the bottom of the stellar convection zone.

An extension of the overshoot models from the Sun to other main-sequence stars has been made by Belvedere et al. (1987). Each star has an overshoot layer which extends over a fraction of the scale height at the base of its convection zone. According to (26) and (27), then, a F star can store only rather weak flux tubes, while a K star may support tubes with fields exceeding 10^5 G in its overshoot layer.

Turbulent transport of magnetic fields has been further investigated by Molchanov et al. (1984), Vainshtein and Kichatinov (1986), and by Hoyng (1987a,b). Hoyng emphasizes the role of the fluctuations, and derives an equation for $\overline{bb}$. This is an important aspect because, as we know, the strength of the magnetic field fluctuations on the Sun exceeds the solar mean field by orders of magnitude.

Linear mean field dynamos in spherical geometry, but without special reference to a magnetic layer as discussed in Sect. 2, have been systematically studied by Rädler (1986), Bräuer and Rädler (1987), and Yoshimura et al. (1984abc). These studies bear on the question of mode selection, e.g. whether a mean field of odd or even parity will be excited, or whether the field is oscillatory or steady. Hoyng (1987bc) considers the stochastic excitation of the various dynamo modes. Observationally, such modes have been analysed by Stenflo and Vogel (1986). Their main result, the appearance of the solar cycle period of 22 years in all odd zonal harmonic contributions, nicely confirms the coupling of these harmonics by means of a symmetric (with respect to the equator) angular velocity, and antisymmetric α-effect, in an αω-dynamo.

Aperiodic phenomena, such as the Maunder minimum in the 17th century, have been attributed to the fact that the mean field dynamo is itself a dynamical system. Non-linear interaction terms, and sufficient degrees of freedom, allow for the desired chaotic behaviour (Weiss et al., 1984), although a different explanation, based on a stationary field in the solar core, has been offered by Pudovkin and Benevolenska (1985). A dynamical system similar to the complex Lorenz system of Weiss et al. has been investigated by Schmalz (1985). He studied the transition to chaos by determining the critical dynamo number for which the first Lyapunov exponent turns positive. As the dimension of the system was increased there are indications of convergence, although, due to the large computational expense, a firm result could not yet be established.

References

Belvedere, G.: 1985, Solar Phys. 100, 363

Belvedere, G., Pidatella, R.M., Stix, M.: 1987, Astron. Astrophys. 177, 183

Bogdan, T.J., Zweibel, E.G.: 1985, Astrophys. J. 298, 867

Bräuer, H.-J., Rädler, K.-H.: 1987, Astron. Nachr. 308, 27

Christensen-Dalsgaard, J., Duvall Jr. T.L., Gough, D.O., Harvey, J.W., Rhodes Jr.,
E.J.: 1985, Nature 315, 378

Deinzer, W., Knölker, M., Voigt, H.H. (eds.): 1986, Small Scale Magnetic Flux
Concentrations in the Solar Photosphere, Proceedings, Vandenhoeck & Ruprecht,
Göttingen

DeLuca, E.E.: 1986, Thesis, University of Colorado, NCAR/CT-104

DeLuca, E.E., Gilman, P.A.: 1986, Geophys. Astrophys. Fluid Dynamics 37, 85

Drummond, I.T., Horgan., R.R.: 1986, J. Fluid Mech. 163, 425

Drummond, I.T., Duane, S., Horgan, R.R.: 1984, J. Fluid Mech. 138, 75

Duvall, T.L., Harvey, J.W.: 1984, Nature 310, 19

Fearn, D.R., Proctor, M.R.E.: 1984, Phys. Earth Planet. Inter. 36, 78

Fossat, E., Gelly, B., Grec, G., Pomerantz, M.: 1987, Astron. Astrophys. 177, L47

Garcia de la Rosa, J.I.: 1987, see Schröter et al. (1987)

Gilman, P.A.: 1983, Astrophys. J. Suppl. Ser. 53, 243

Gilman, P.A., Miller, J.: 1981, Astrophys. J. Suppl. Ser. 46, 211

Gilman, P.A., Miller, J.: 1986, Astrophys. J. Suppl. Ser. 61, 585

Glatzmaier, G.A.: 1984, J. Comp. Phys. 55, 461

Glatzmaier, G.A.: 1985a, Astrophys. J. 291, 300

Glatzmaier, G.A.: 1985b, Geophys. Astrophys. Fluid Dynamics 31, 137

Glatzmaier, G.A.: 1987, in The Internal Solar Angular Velocity (B.R. Durney, S.
Sofia, eds.), Reidel, in the press

Gough, D.O.: 1987, in The Internal Solar Angular Velocity (B.R. Durney, S. Sofia,
eds.), Reidel, in the press

Hoyng, P.: 1985, J. Fluid Mech. 151, 295

Hoyng, P.: 1987a, Astron. Astrophys. 171, 348

Hoyng, P.: 1987b, Astron. Astrophys. 171, 357

Hoyng, P.: 1987c, Nature, submitted for publication

Isaak, G.R., Jefferies, S.M., McLeod, C.P., New, R., van der Raay, H.B., Pallé,
P.L., Régulo, C., Roca Cortés, T.: 1987, in Advances in Helio- and
Astroseismology (J. Christensen-Dalsgaard, ed.), IAU Symp. 123, Reidel,
in the press

Knobloch, E.: 1977, J. Fluid Mech. 83, 129

Knobloch, E.: 1978a, Astrophys. J. 220, 330

Knobloch, E.: 1978b, Astrophys. J. **225**, 1050

Kraichnan, R.H.: 1976, J. Fluid Mech. **77**, 753

Krause, F.: 1967, Habilitationsschrift, Univ. Jena

Krause, F.: 1968, Z.A.M.M. **5**, 333

Krause, F., Rädler, K.-H.: 1980, Mean Field Magnetohydrodynamics and Dynamo Theory,
    Pergamon Press, Oxford

Krivodubskii, V.N.: 1984a, Astron. Zh. **61**, 354

Krivodubskii, V.N.: 1984b, Astron. Zh. **61**, 540

Leighton, R.B.: 1964, Astrophys. J. **140**, 1547

Moffatt, H.K.: 1974, J. Fluid. Mech. **65**, 1

Moffatt, H.K.: 1978, Magnetic Field Generation in Electrically Conducting Fluids,
    Cambridge Univ. Press

Molchanov, S.A., Ruzmaikin, A.A., Sokoloff, D.D.: 1984, Geophys. Astrophys. Fluid
    Dyn. **30**, 242

Moreno-Insertis, F.: 1983, Astron. Astrophys. **122**, 241

Moss, D: 1986, Phys. Rep. **140**, 1

Nicklaus, B.: 1987, Diplomarbeit, Univ. Freiburg

Noyes, R.W., Weiss, N.O., Vaughan, A.H.: 1984, Astrophys. J. **287**, 769

Pallé,P.L., Pérez, J.C., Régulo, C., Roca Cortés, T., Isaak, G.R., McLeod,
    C.P., van der Raay, H.B.: 1986, Astron. Astrophys. **170**, 114

Parker, E.N.: 1955a, Astrophys. J. **121**, 491

Parker, E.N.: 1955b, Astrophys. J. **122**, 293

Parker, E.N.: 1975, Astrophys. J. **198**, 205

Pidatella, R.M., Stix, M.: 1986, Astron. Astrophys. **157**, 338

Pudovkin, M.I., Benevolenska, E.E.: 1985, Solar Phys. **95**, 381

Rädler, K.-H.: 1986, Astron. Nachr. **307**, 89

Roberts, B., Campbell, W.R.: 1986, Nature. **323**, 603

Rosner, R., Weiss, N.O.: 1985, Nature **317**, 790

Schmalz, S.: 1985, Diplomarbeit, Univ. Freiburg

Schmidt, H.U. (ed.): 1985, Theoretical Problems in High Resolution Solar Physics,
    Proceedings, MPA 212, München

Schmitt, D.: 1984 in The Hydromagnetics of the Sun (P. Hoyng, ed.), Proceedings,
    ESA SP-220, p. 223

Schmitt, D.: 1985, Thesis, Univ. Göttingen

Schmitt, D.: 1987, Astron. Astrophys. **174**, 281

Schmitt, J.H.M.M., Rosner, R., Bohn, H.U.: 1984, Astrophys. J. **282**, 316

Schröter, E.H., Vazquez, M., Wyller, A. (eds.): 1987, The Role of Fine-Scale
    Magnetic Fields on the Structure of the Solar Atmosphere, Proceedings,
    Cambridge Univ. Press, Cambridge

Schüssler, M.: 1977, Astron. Astrophys. **56**, 439

Schüssler, M: 1983, in _Solar and Stellar Magnetic Fields: Origins and Coronal Effects_ (J.O. Stenflo, ed.), IAU Symp. **102**, p. 213

Shaviv, G., Salpeter, E.E.: 1973, _Astrophys. J._ **184**, 191

Spiegel, E.A., Weiss, N.O.: 1980, _Nature_ **287**, 616

Spruit, H.C., van Ballegooijen, A.A.: 1982, _Astron. Astrophys._ **106**, 58

Steenbeck, M., Krause, F.: 1969, _Astron. Nachr._ **291**, 49

Stix, M.: 1976, _Astron. Astrophys._ **47**, 243

Stix, M.: 1984, _Astron. Nachr._ **305**, 215

Unno, W., Ribes, E.: _Astrophys. J._ **208**, 222

Vainshtein, S.I., Kichatinov, L.L.: 1986, _J. Fluid Mech._ **168**, 73

van Ballegooijen, A.A.: 1982, _Astron. Astrophys_ **113**, 99

van Ballegooijen, A.A. (1986): unpublished manuscript

Vorontsov, S.V.: 1987, in _Advances in Helio- and Astroseismology_ (J. Christensen-Dalsgaard, ed.), IAU Symp. **123**, Reidel, in the press

Weiss, N.O.: 1986, in _Highlights of Astronomy_ (J.-P. Swings, ed.), Vol. 7, p. 385

Weiss, N.O.: 1987, in _Physical Processes in Comets, Stars and Active Galaxies_ (W. Hillebrandt, E. Meyer-Hofmeister, H.-C. Thomas, eds.), Proceedings, Springer, p. 46

Weiss, N.O., Cattaneo, F., Jones, C.A.: 1984, _Geophys. Astrophys. Fluid Dyn._ **30**, 305

Woodard, M.F., Noyes, R.W.: 1985, _Nature_ **318**, 449

Yoshimura, H.: 1972, _Astrophys. J._ **178**, 863

Yoshimura, H.: 1976, _Solar Phys._ **50**, 3

Yoshimura, H., Wang, Z., Wu, F.: 1984a, _Astrophys. J._ **280**, 865

Yoshimura, H., Wang, Z., Wu, F.: 1984b, _Astrophys. J._ **283**, 870

Yoshimura, H., Wu, F., Wang, Z.: 1984c, _Astrophys. J._ **285**, 325

Zweibel, E.G., Bogdan, T.J.: 1986, _Astrophys. J._ **308**, 401

STELLAR ACTIVITY AND ROTATION

Marcello Rodonò

Astronomical Institute of Catania University
and Astrophysical Observatory
Viale Andrea Doria 6, I-95125 Catania, Italy

ABSTRACT

The most important signatures and parameters of stellar magnetic activity and their relation with global stellar parameters are briefly reviewed with the aim of indicating which are the most significant data that are required to constrain possible models of stellar activity. The sporadic and cyclic variability aspect is particularly stressed, firstly, because it is a crucial activity parameter and, secondly, because of its possible effects on the derived general correlations between activity and stellar parameters. In particular, the rotation rate, although it is recognized to play a relevant role, still its correlation with various activity indicators has an essentially qualitative character, because other global stellar parameters undergo concurrent changes, making it difficult to isolate the pure effect of stellar rotation from observations.

1. INTRODUCTION

In order to comply with the Scientific Organizing Committee's recommendation to outline problem areas, rather than established results, I will only briefly summarize the most important signatures of stellar activity for the purpose of attempting a working definition of "stellar activity" and "activity level", and to indicate what the present observational limits allow us to say about the most relevant parameters characterizing stellar activity phenomena. To this purpose I shall consider only those phenomena of variability, which may be ascribed, from direct or circumstantial evidence, to the release of magnetic energy. Threfore only magnetic activity phenomena will be considered even if, for the sake of brevity, I will often refer to them simply as activity.

Taking advantage of our detailed knowledge, though not complete understanding of solar activity, we may presume that hypothetical observations of the Sun at stellar distance would show long term variability of chromospheric, transition region and coronal spectral diagnostics. This variability may be correctly interpreted in terms of plage and coronal feature changes associated with the 11-years' activity cycle. Moreover, sporadic variability, due to the formation and decay of active regions, should be expected.

Therefore, Wilson's programme of monitoring stellar Ca II H and K line fluxes was based on well established scientific footings, as successfully demonstrated after a dozen year efforts (Wilson 1978, Baliunas et al. 1983). Similarly, the measurements of global emission from stellar chromospheres, which were pioneered by Otto Struve, or from transition regions (Linsky et al. 1978), and coronae (Vaiana et al. 1981) are suggestive of solar-like chromospheric structures. Also,

short-term flare-like variability on red-dwarf stars has been
successfully investigated since the early sixties (cf. Gershberg 1970
and references therein).

Instead, a straightforward extension of the solar-stellar analogy to
photospheric activity phenomena should have discouraged from the start
any search for stellar variability due to sunspot-like inhomogeneities: the remarkable variability of sunspot number and dimensions
during the course of the 11-year solar cycle would not be detectable
at stellar distance. Only occasionally, large sunspot complexes would
give rise to less than one per cent variability of the the global
photospheric output, i.e. at the limit of the present photometric
precision. Therefore, a bold and a priori unsupported extrapolation of
the solar-stellar analogy was behind Kron's (1950) suggestion that
the low-amplitude ($\approx$ 0.1 mag) wide-band photometric variations, which
were observed in a few K-M emission-line dwarfs and subgiants, might
be attributed to huge solar-like spots, whose visibility was modulated
by the stars' rotation. Spotted areas, covering 10-40 per cent of the
projected stellar disk, and spot temperatures cooler than the
surrounding photosphere by about 300 to 1500 degrees have been derived
from modelling the most recent and accurate rotation-induced modulation of wide-band optical flux (cf. Vogt 1983, Rodonò 1986a).
Therefore, stellar rotation first enters the field of stellar activity
as a passive device to modulate the variable aspects of starspots, as
they move across the projected stellar disk. This circumstance,
although merely incidental for the stellar activity case, need not to
be underestimated. Firstly, because it offers a basic method to derive
quantitative information on the physical characteristics of starspot
and plages, and secondly, because it allows us to measure the rotation
period of stars with an accuracy comparable or even better than
spectroscopic observations. Actually, stellar rotation was soon
recognized to play a more significant role because active stars were
identified as relatively fast rotators with v sin i > 5 Km/s (Bopp
and Fekel, 1977) and rotation turned out to be a key ingredient in
producing and reinforcing stellar magnetic fields via dynamo
mechanism, i.e. in providing one basic agent for the development of
stellar magnetic activity (Parker 1959, 1986; Belvedere 1983; Gilman
1983; Schussler 1983, and references therein).

2. STELLAR MAGNETIC ACTIVITY: A WORKING DEFINITION

As already stressed in the Introduction, magnetic fields and intrinsic
variability are the characteristic and identifying aspects of those
stellar activity phenomena, which are collectively referred to as
magnetic activity. Therefore, magnetic A-type stars, which show
intrinsically stable magnetic fields and luminosity, are not active in
the sense outlined above.

The intrinsic variability of active stars is presumably linked
to the variability of magnetic field topology and strength. The
emergence of magnetic flux tubes above the photosphere, from deep in
the convection zone, and their extension up to coronal levels, in the
form of closed or open field structures, lead to the production of
sunspot-like features in the photosphere, plages in the chromosphere
and transition region and coronal heathing. Observational evidence of
spatial correlations between localized surface inhomogeneities on
stars at various atmospheric levels is shown in Figure 1: photospheric
spots on II Pegasi are overlain by chromospheric and transition plages

(Rodonò et al. 1987), as on the Sun. This result puts on firm
observational footings the concept that homogeneous stellar
atmospheres are only first order theoretical idealizations. The real
situation implies, as observed, intrinsic variability on time scales
appropriate to the decay times of such localized magnetic structures.
Therefore, the complex atmospheric structure of stellar atmospheres
and their intrinsic variability are the two basic requirements that
should underly any working definition of stellar activity. Bearing in
mind these two basic requirements, an active star is characterized by:

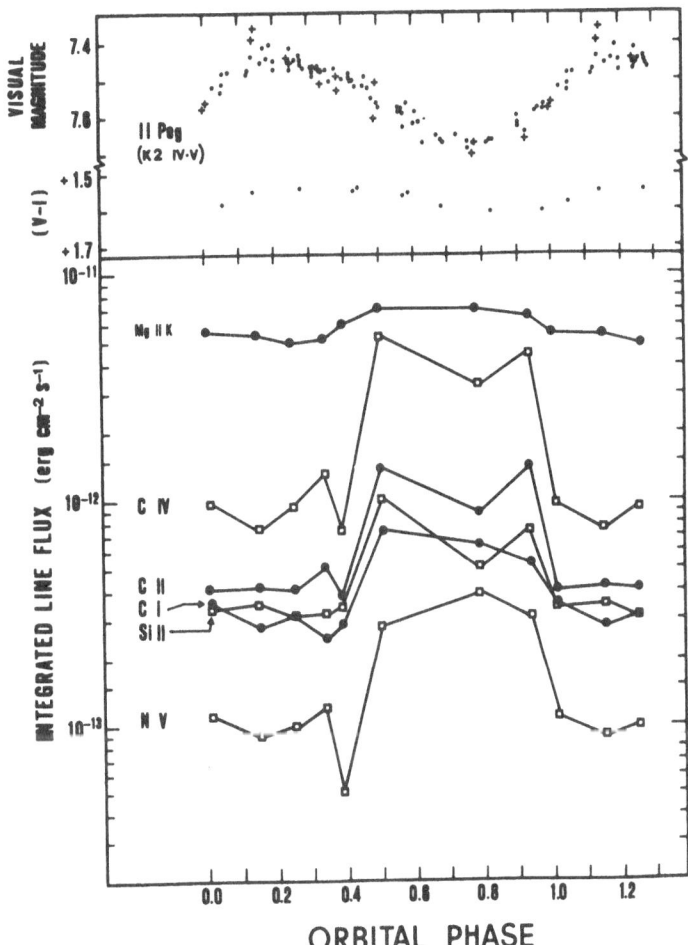

Figure 1. Rotation-induced modulations of V-band, V-I color (top
panel) and integrated flux in major emission lines (bottom panel) for
II Pegasi. Maximum photospheric cool spot visibility, i.e. the light
minimum, appears to correlate with maximum chromospheric and
transition region bright plage visibility, i.e. with line flux maximum
(adapted from Rodonò et al. 1987).

a) magnetic fields with complex topology, as implied by,

b) **surface inhomogeneities**, which give rise to rotation-induced
modulation of continuum and line fluxes on time scales of a few
days (Figure 1);

c) intrinsic variability on time scales of the order of 10 days,
due to the evolution and decay of the localized magnetic
structures, and of the order of 1-10 years, as a result of long
term activity cycles (Figures 2 and 3);

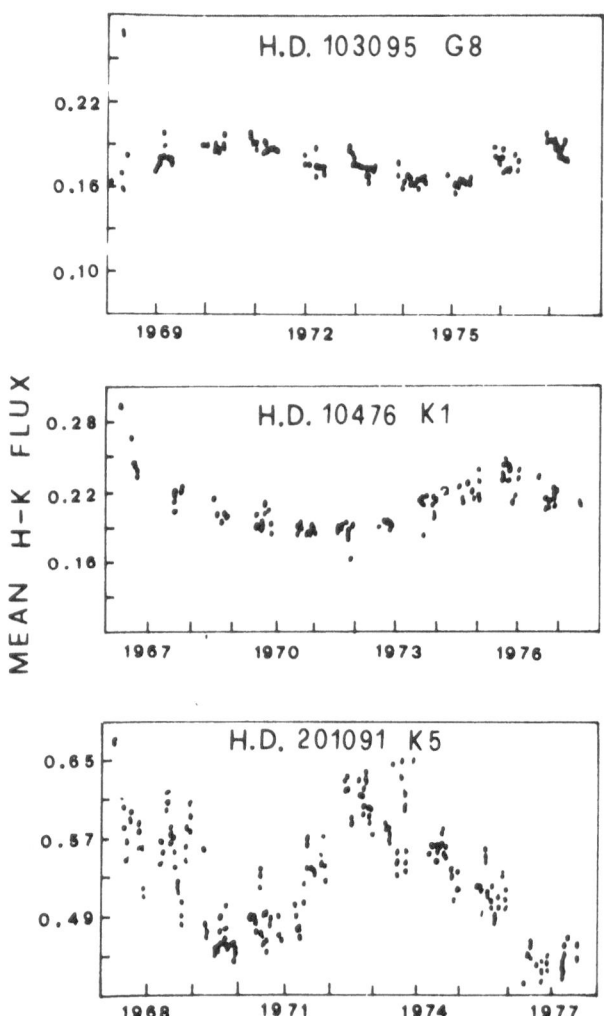

Figure 2. Long-term cyclic variation of Ca II H and K line flux for
selected stars from Wilson's (1978) monitoring program. The large
spread of the bottom plot is likely due to rotation-induced modulation
of the observed flux.

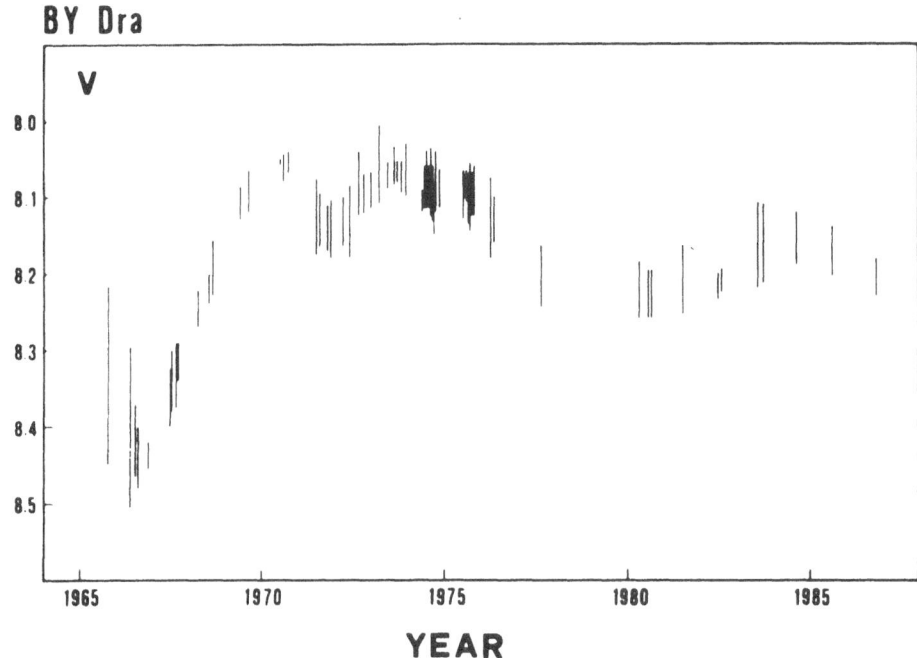

Figure 3. Long-term V-band cyclic variation of BY Draconis from a recollection of recent photometry from several sources (Cutispoto 1987). The length of vertical bars indicates the amplitude of the 3.836 day rotation-induced flux modulation attributable to cool spots.

    d) intrinsic variability on time scales of 1-10 minutes, due to flare-like events, which may affect almost simultaneously the entire stellar atmosphere (Figure 4a and 4b).

As far as activity level is concerned, I should like to stress that, due to the long term character of stellar activity, as outlined above, the term level should imply the existence of an activity cycle and, consequently, a variable level of activity. If so, the characteristic parameters of what we call activity level are i) the cycle phase, ii) the mean emission flux from all active areas integrated over the star's surface and iii) the surface covering factor or filling factor relative to a given parameter. Often the term activity level is used as synonimous of "total emission", from where the qualitative attributes of "low level" or "high level" emitting stars. Generally, this terminology is the consequence of missing evidence on the possible existence of an activity cycle, so that the observed "total emission" is referred to as the star's "activity level", without any reference to its possible time variability.

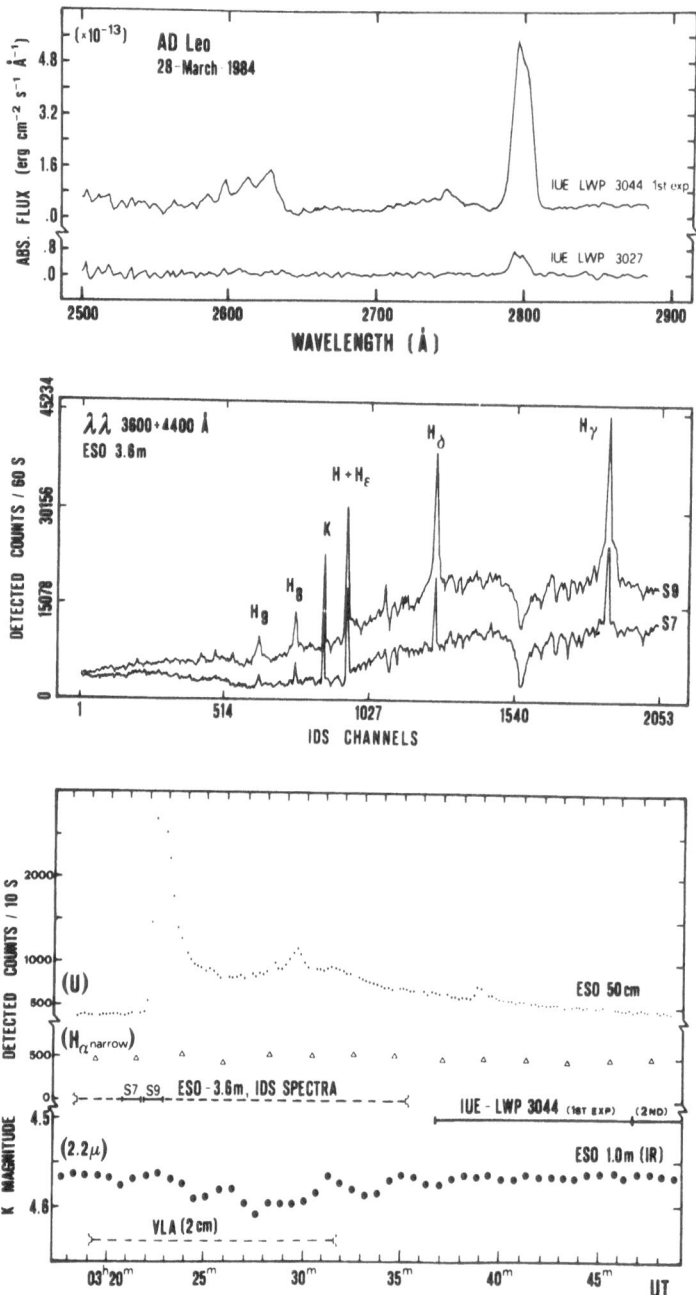

Figure 4a. The 1984, March 28 flare of AD Leo observed simultaneously with IUE (upper panel), the Image Dissector Scanner at the ESO 3.6-m telescope (middle panel), optical and IR photometers (lower panel) at the ESO 50-cm and 1.0-m telescopes (from Rodonò et al 1984).

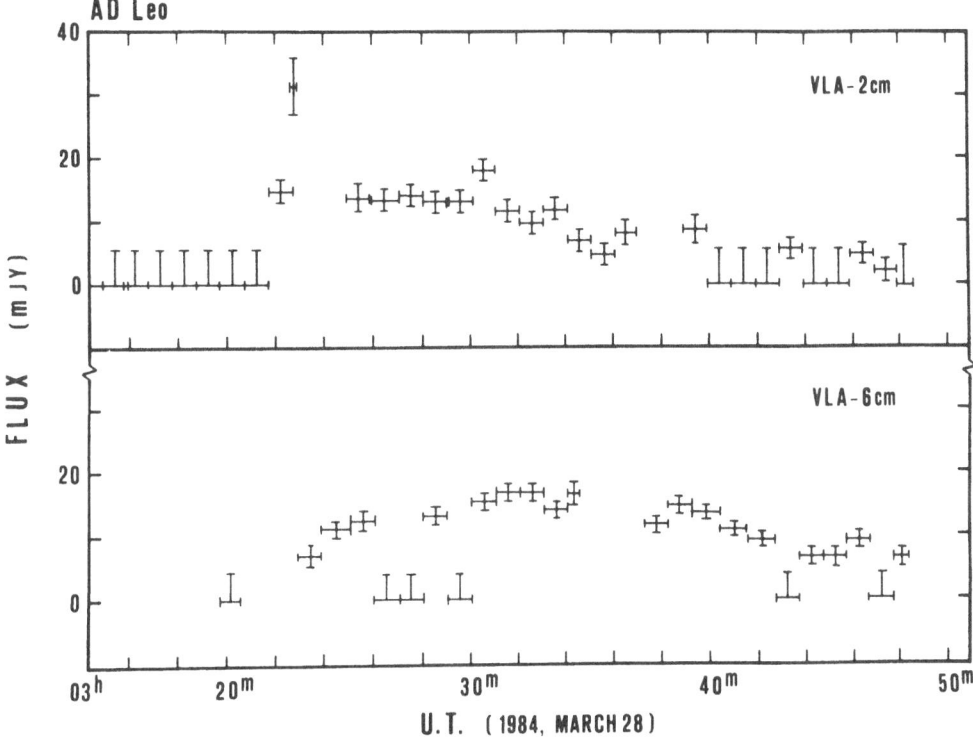

Figure 4b. Simultaneous 2- and 6-cm VLA observations of the AD Leo
flare in Figure 4a obtained by D.E.Gary and D.M.Gibson (cf. Rodonò
1986c).

3. ACTIVITY VERSUS GLOBAL STELLAR PARAMETERS

In order to constraint the possible mechanisms and models of stellar
activity it is considered extremely important to search for
quantitative functional correlations between activity indicators, and

   a) global stellar parameters, such as mass, luminosity, effective
   temperature, surface gravity, age (or chemical composition), depth
   of convection zone, rotation regime, and magnetic fields,

   b) the physical characteristics of the plasma, such as temperature,
   density and, possibly local magnetic field, as well as the active
   region extension, covering factor and location in the atmosphere.

The diagnostic tools of stellar activity are numerous as well. The
most used ones are: the continuum and line flux variability and line
profile changes due to more or less compact active areas on the star's
atmosphere. The number of parameters is large enough to allow us a

number of exercises, but it is difficult to identify the most meaningful combination of parameters for the stellar activity case. From the theoretical point of view, the most meaningful ones are (differential) rotation, depth of convection zone, the resulting magnetic field and filling factor, i.e. those parameters which constitute the recipe of α-ω dynamo mechanism. From the observational point of view, the situation is complicated by the mutual dependence of global stellar parameters among each other, that makes very difficult to disentangle the effect of individual crucial parameters.

Several more or less definite quantitative correlations between activity indicators and global stellar parameters can be found in the literature (cf. review papers in Byrne and Rodonò 1983, Stenflo 1983, Mangeney and Praderie 1984b, Zeilik and Gibson 1986). Since it is beyond the aim of the present paper to review all of them, only a brief outline of major results follows:

a) The Ca II H and K flux (F'_{HK}), normalized to the bolometric flux (σT^4_{eff}), depends on the magnetic field strength (B) and on its filling factor, according to the following relation (Marcy 1983):

$$F'_{HK}/\sigma T^4_{eff} = 6.14 \times 10^{-14} \times B^{0.5} \times T^2_{eff} \times f^{0.6} \qquad (1)$$

More accurate and extended data on surface magnetic fields (cf. §e) may require in the future some revision of the above relation, but the interdependence of the above quoted parameters appears to be well established.

b) Various indicators of activity, such as the absolute luminosity in the chromospheric Ca II H and K emission lines (L_{HK}), in the transition region C IV line (L_{CIV}) and in the coronal X-ray emission (L_x), show the following power law dependence on angular velocity:

$$L_{HK} \approx f_{chr}(M/Mo) \times \Omega$$

$$L_{CIV} \approx f_{TR}(M/Mo) \times \Omega^{1.5} \qquad (2)$$

$$L_x \approx f_{cor}(M/Mo) \times \Omega^{2+3}$$

or exponential dependence on rotation period (P):

$$L_{HK} \times f'_{chr}(M/Mo) \approx 10^{-(P/31.7)}$$

$$L_{CIV} \times f'_{TR}(M/Mo) \approx 10^{-(P/22.5)} \qquad (3)$$

$$L_x \times f'_{cor}(M/Mo) \approx 10^{-(P/10.4)}$$

where f(M/Mo) are mass or B-V dependent functions along the main sequence (Catalano 1984, Marilli et al. 1986, and references therein). Moreover, Marilli et al. (1986) have shown that in the first of correlations (3), on a log-log scale,

$$\log L_{HK} = b - a \times P \qquad (4)$$

both a and b coefficients are color dependent (Figure 5).

c) The ratio of chromospheric flux to total bolometric flux in the Ca II H and K emission line (R'HK) is well correlated with the parameter Ro = P/τ_c(B-V) (Rossby number), where τ_c(B-V) is the convective turnover time, calculated assuming a mixing length to scale height ratio $\alpha = 2$ (Noyes et al. 1984). The authors consider their correlation more consonant with the general predictions of dynamo theory, which relates the magnetic field generation, and consequently the magnetic activity, to both rotation rate and depth of convection zone. They claim that the parameters R'_{HK} and Ro appear more tightly correlated with the rotation period P than L_{HK} or F_{HK} are. However, Basri (1986) has demonstrated that there is no reason, at least on the basis of the observational scatter, to prefer one set of variables over the other, because the shape of the relation is largely preserved in both cases (Figure 6). On the other hand, Mangeney and Praderie (1984a) have shown that the X-ray luminosity (L_x) and flux (F_x) are very well correlated with both the angular velocity (Ω) and the total depth of convection zone (L_c), via an "effective" Rossby number $R^* = \frac{1}{2} Vm/(\Omega \times L_c)$, where Vm is the maximum convective velocity:

$$L_x \quad \alpha \quad R^{*1/2}$$

$$F_x \quad \alpha \quad R^{*3.3} \tag{5}$$

The question whether a convective parameter, such as τ_c or R^*, should be taken into account, together with rotation, in activity versus global stellar parameter correlations is still open. Rutten and Schrijver (1986) suggest that the best parameters to be used are i) the flux excess (F_{HK}), above an observationally determined minimal flux, and, as Marilli et al. (1986), ii) the simple rotation period (P) or angular velocity (Ω). The rationale of this choice goes in the direction of not considering combinations of concurrently changing parameters, unless necessary (cf. § 4).

d) The indicators of global activity on dwarfs and giants at different atmospheric levels show power law correlations (Ayres et al. 1981, Schrijver 1983, Mangeney and Praderie 1984a), which are consistent with their different dependence on rotation (Catalano 1984):

$$L_{CIV} \quad \approx \quad L_K^{1.8} \, g_{TR}(M/Mo)$$

$$L_x \quad \approx \quad L_K^{2.8+3} \, g_{cor}(M/Mo) \tag{6}$$

$$F_x \quad \alpha \quad F_{HK}^{1.7}$$

where the g(M/Mo) functions indicate some dependence on mass. The latter relation (Schrijver 1983) is based on a multidimensional analysis including a more extensive sample of dwarfs and giants than ever before. Correlations (6) indicates that non-thermal energy deposition in stellar atmospheres increases with height, i.e. with temperature. This trend of the global emission appears to be qualitatively similar for solar and stellar active areas, but the latter show progréssively larger energy losses, up to two orders of magnitude than solar, at high atmospheric levels (Rodonò 1983).

e) From Robinson-type Zeeman analysis (Robinson 1980, 1986; Marcy 1983, 1984) of a number of lines with different Landè g factors, and taking into account radiative transfer effects and compensation

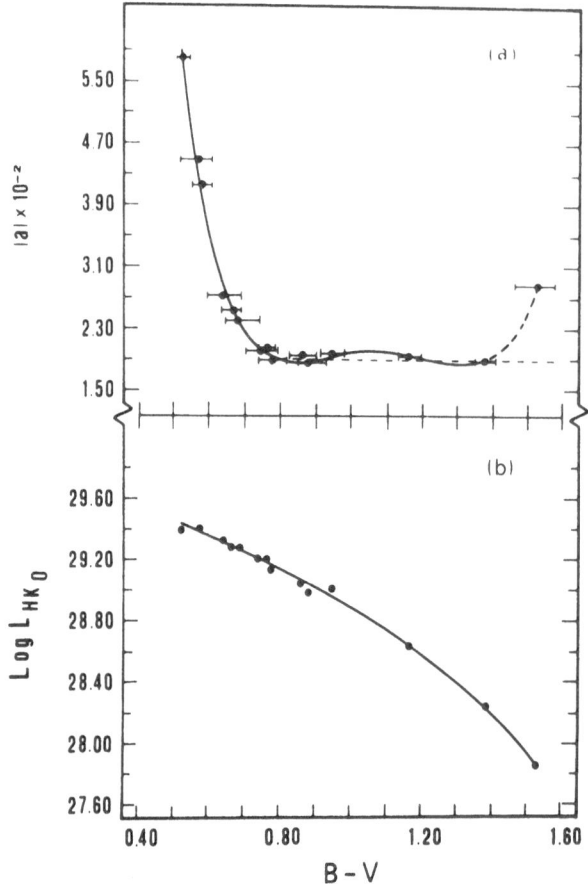

Figure 5. The trend of a and b coefficients in relation (4) versus B-V (from Marilli et al. 1986) indicating that the period-activity relation for late type stars depends on spectral type or mass.

for blends, Saar et al. (1985) and Saar and Linsky (1987) were able to derive reliable values of magnetic field strenghts, ranging from 1000 to 3800 Gauss (Figure 7), for about two dozen dwarf stars of spectral types G0 (/' Ori) to M3.5e (AD Leo). The magnetic field strenght appears to increase with B-V and g, i.e. towards later spectral types, and the filling factors increase with angular velocity, as suggested by dynamo theory (Durney and Robinson 1982). By taking the magnetic pressure equal to the gas pressure (Pg), the resulting magnetic fields B_{eq} [$= (8\pi Pg)^{1/2}$] are tightly correlated with photospheic magnetic field strenghts, suggesting that magnetic flux tubes are confined by photospheric gas pressure. Accurate measurements of stellar magnetic fields must be considered a high priority target of stellar activity studies for several years to come.

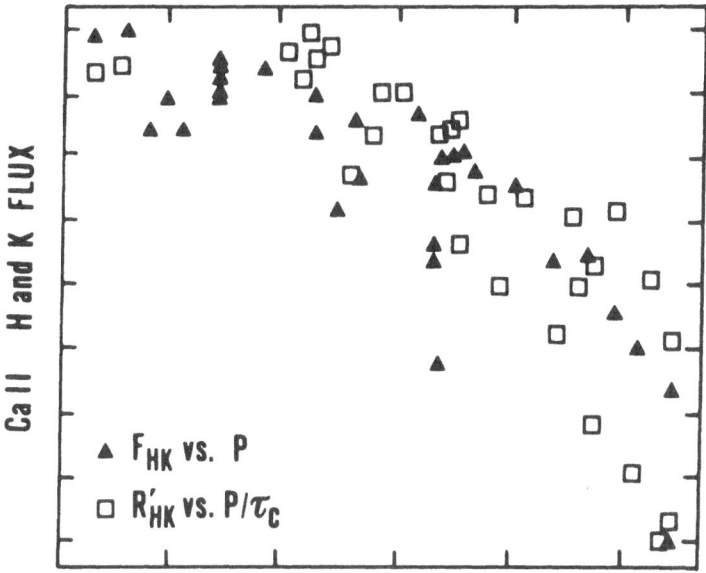

Figure 6. Ca II H and K line surface flux F_{HK} versus rotation period (filled symbols) and flux normalized to the bolometric one R'_{HK} versus Rossby number, P/τ_c (open symbols). The plots are normalized to the range covered by each set of variables. Observational scatter does not allow us to prefer one set of variables over the other.

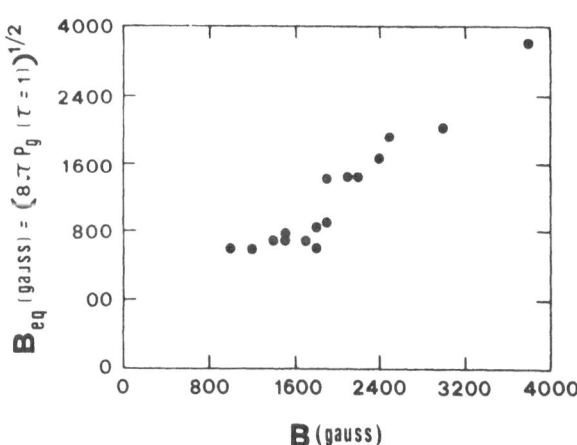

Figure 7. Equipartition (Beq) versus measured (B) magnetic field strengths suggesting that photospheric gas pressure confines magnetic flux tubes on late type stars and determines their photospheric magnetic field strengths (Saar and Linsky 1987).

4. ACTIVITY VERSUS ROTATION: MACRO OR MICRO-CORRELATIONS?

The choice of the title for my talk by the Scientific Organizing Committee may imply that rotation is the most relevant parameters of stellar activity. Rotation is certainly important provided that it is instrumental in generating and maintaining a strong and complex magnetic field, which, eventually, is the real triggering agent of stellar activity. In other words, fast rotation not coupled with sufficiently vigorous convection and surface turbulence might be of null interest for the stellar activity case.

In the preceeding section some well established correlations between activity indicators and rotation for main sequence and giant stars were presented. These correlations are certainly important because clearly indicate the relevant effect of rotation on activity, in qualitative agreement with the general prediction of dynamo theory. However, we must bear in mind that in the H-R diagram and, particular-ly, along the main sequence, the luminosity, mass, radius, color index, line spectrum, age, rotation itself, Rossby and dynamo numbers, and other parameters, all change simultaneously, monotonically, and are mutually interdependent. Therefore, the search for quantitative correlations among activity indicators and rotation, which aim at covering a wide range of stellar types and ages, may lead to quanti-tatively inconclusive or even misleading results, especially when the validity of the resulting correlations relays only on the criterion of the smallest dispersion of data points about a given trend. For instance, it is indicative that, while the UV emission of active chromosphere stars decreases in strength with declining rotation, as found for dwarf stars, there is no dependence of emission upon rota-tion within individual luminosity classes and no direct support for a Rossby-type rotation-activity relation for the highly active RS CVn binaries, as an isolated group, is apparent (Simon 1986). This result suggests that, at high emission levels, other secondary parameters or possibly non linear effects may dominate the mechanism that leads to the observed activity. As a matter of fact, individual stars may behave uniquely, as demonstrated by the observations of quite diffe-rent activity levels on stars, which appear similar in all other respects.

In close binaries, as RS CVn or BY Dra systems, tidal coupling may well affect their rotation regime up to some depth into the stars, with obvious consequences in their surface activity manifestation. The fact that RS CVn binaries, as a group, follow the general rotation-activity relation for dwarfs and giants, is suggestive of a dynamo operating at the bottom of the convection zone, deep into the star interior, where the rotation regime may remain relatively unaffected by tidal coupling.

Therefore, in order to make some progress in our understanding of stellar activity, detailed studies of specifically selected stars are needed, rather than seeking for, or trying to improve established correlations between activity and global stellar parameters spanning over a wide range of stellar types. Any such correlation might be deceptive if concurrent effects of several parameters are not clearly singled out.

In view of the complexity of non-linear effects affecting the development of stellar activity phenomena, firm constraints on theory may be provided only by accurate, systematic and synotpic observations. The principal parameters of interest to contraint theory, or suggested by the present theoretical models, are:

a) chromospheric, transition region and coronal energy losses from
simultaneous ground-based and space observations (Rodonò et al.
1984, Butler et al. 1986, Doyle et al. 1986, Monsignori Fossi et
al. 1986, Rodonò et al. 1987, Byrne et al. 1987, Haisch et al
1987);

b) rotation velocity from line profile fittings and wide-band
photometric variations (Baliunas et al. 1983, Stauffer and Hartmann
et al. 1987, Stauffer et al. 1987);

c) activity cycle period and amplitude (Phillips and Hartmann 1978,
Rodonò et al. 1983, Mavridis et al. 1986);

d) size, filling factor, location and migration of surface
structures at different atmospheric levels from rotational
modulation of continuum and integrated line fluxes (cf. Rodonò
1986a);

e) surface differential rotation (Rodonò 1986b);

f) depth of convection zone from line bisector (cf. Dravins 1987)
and oscillation studies (cf. Christensen-Dalsgraard 1986), which
may throw some light on radial differential rotation;

g) last, but far from least, photospheric magnetic field strength
and filling factors, from comparying observed high spectral resolution and theoretical line profiles of magnetic sensitive and
insensitive lines to measure their relative Zeeman splitting
(Robinson 1986, Saar et al. 1985, Saar and Linsky 1985).

Only some relevant and recent papers, dealing with the general
methodology to derive the appropriate parameters and containing data
recollection or new determinations, are quoted above, because it is
not viable and beyond the purpose of the present paper to review all
of them.

A huge amount of data will be obviously required to address the above
listed points, especially because one basic information, to be
provided by synoptic observations, is the variability time scales of
activity parameters. Variability is often disregarded in general
correlation studies or, even worse, is improperly invoked as a "deus
ex machina" to account for unpleasantly displaced data or for their
large dispersion about the general trend. We need to relay on observations with internal accuracy of the order of a few thousands per cent
and, for oscillation studies, at least two orders of magnitude better.
If calibrations, transformations or normalizations are adopted, it is
very important that also the original observations are presented.

In order to collect the huge amount of uniform, consistent and
systematic multi-band data that is needed, the use of automated telescopes (cf. Genet et al. 1987) and the implementation of international
networks of dedicated telescopes (cf. Giampapa 1986), together with
simultaneous UV and X-ray observations, will prove to be essential.

5. CONCLUSION

The main conclusion that can be drawn from the above arguments is that a new research attitude is necessary to progress significantly in the study of stellar activity: instead of macro-correlations, which involve activity and global parameters' covering a wide range of star types, we need to search for micro-correlations between one single varying parameter as a function of another one. This can be done only if even limited samples of stars, similar in all other respects except the two parameters of interest, are sorted out from the variety of stellar types that nature offers to our consideration.

Acknowledgments.

I should like to thank the Scientific and Local Organizing Committees for inviting me to attend a successful meeting at Titisee and for generous hospitality. The Catania Astrophysical Observatory is also gratefully acknowledged for complementary fund allocation. The research program on stellar activity at Catania is supported by the CNR: Gruppo Nazionale di Astronomia and by the Ministero della Pubblica Istruzione of Italy.

REFERENCES

Ayres T.R., Marstad N.C., Linsky J.L.: 1981, Astrophys. J. 247, 545.
Baliunas S.L., Vaughan A.H., Hartmann L., Middlekoop F., Mihalas D., Noyes R.W., Preston G.W., Frazer J., Lanning H: 1983, Astrophys.J. 275, 752.
Basri G.: 1986, in Cool Stars, Stellar Systems, and the Sun, M. Zeilik and D.M.Gibson eds., Springer-Verlag, Berlin, p. 184.
Belvedere G.: 1983, in IAU Coll. 71, P.B.Byrne and M.Rodonò eds., Reidel Publ. Co., Dordrecht, p. 579.
Bopp B.W., Fekel F.: 1977, Astron. J. 82, 490.
Butler C.J., Rodonò M., Foing B.H., Haisch B.M.: Nature 321, 679.
Byrne P.B., Rodonò M. (eds.): 1983, Activity in Red-Dwarf Stars, IAU Coll. 71, Reidel Publ. Co., Dordrecht.
Byrne P.B., Doyle J.G., Brown A., Linsky J.L., Rodonò M.: 1987, Astron. Astrophys., in press.
Catalano S.: 1984, in Space Research Prospects in Stellar Activity and Variability, A.Mangeney and F.Praderie eds., Obs. Paris-Meudon Press, p. 243.
Christensen-Dalsgaard J.: 1986, in Cool Stars, Stellar Systems, and the Sun, M.Zeilik and D.M.Gibson eds., Springer-Verlag, Berlin, p. 145.
Cutispoto G.: 1987, Ph.D. Thesis, University of Catania.
Doyle J.G., Butler C.J., Haisch B.M.: 1986, Mon.Not.Roy.Astr.Soc. 223, 1p.
Dravins D.: 1987, These Proceedings.
Durney B.R., Robinson R.D.: 1982, Astrophys. J. 253, 290.
Genet R.M., Boyd L.J., Kissel K.E., Crawford D.L., Hall D.S., Hayes D.S., Baliunas S.L.: 1987, Publ.Astr.Soc.Pacific, in press.
Gershberg R.E.: 1970, Astrophysics 6, 92.
Giampapa M. (ed.): 1987, The SHIRSOG Workshop, Nat. Solar Obs., Nat. Opt. Astr. Obs., Tucson, Arizona.
Gilman P.A.: 1983, IAU Symp. 102, J.O.Stenflo ed., Reidel Publ. Co., Dordrecht, p. 247.

Haisch B.M, Butler C.J., Doyle J.G., Rodonò M.: 1987, **Astron. Astrophys.**, in press.
Kron G.E.: 1950, **Astron. J. 55**, 69.
Linsky J.L., Ayres T.R., Basri G.S., Morrison N.D., Boggess A., Schiffer III F.M., Holm A., Cassatella A., Heck A., Macchetto F., Stickland D., Wilson R., Blanco C., Dupree A.K., Jordan C., Wing R.F.: 1978, **Nature 275**, 389.
Mangeney A., Praderie F.: 1984a, **Astron. Astrophys. 130**, 143.
Mangeney A., Praderie F. (eds.): 1984b, **Space Research Prospects in Stellar Activity and Variability**, Obs. Paris-Meudon Press.
Marcy G.W.: 1983, IAU Symp. 102, J.O. Stenflo ed., Reidel Publ. Co., Dordrecht, p. 3.
Marcy G.W.: 1984, **Astrophys. J. 276**, 286.
Marilli E., Catalano S., Trigilio C.: 1986, **Astron. Astrophys. 167**, 297.
Mavridis L.N., Asteriadis G., Mahmoud F.M.: 1986, in Compendium in Astronomy, Reidel Publ. Co., Dordrecht, p. 253.
Monsignori Fossi B., Landini M., Pallavicini R., Tribioli F.: 1986, XXVI COSPAR, Adv.Space Sci., in press.
Noyes R.W., Hartmann L.W., Baliunas S.L., Duncan D.K., Vaughan A.H.: 1984, **Astrophys. J. 279**, 763.
Parker E.N.: 1979, **Cosmical Magnetic Fields: Their Origin and Their Activity**, Clarendon Press, Oxford.
Parker E.N.: 1986, in Cool Stars, Stellar Systems, and the Sun, M. Zeilik and D.M.Gibson eds., Springer-Verlag, Berlin, p. 341.
Phillips M.J., Hartmann L.: 1978, **Astrophys. J. 224**, 182.
Robinson R.D.: 1980, **Astrophys. J. 239**, 961.
Robinson R.D.: 1986, in Highlights of Astronomy, J.P. Swings ed., Reidel Publ. Co., Dordrecht, p. 417.
Rodonò M.: 1983, **Adv. Space Res. 9**, no.2, 225.
Rodonò M.: 1986a, in Highlights of Astronomy, J.P.Swings ed., Reidel Publ. Co., Dordrecht, p. 429.
Rodonò M.: 1986b, in Cool Stars, Stellar Systems, and the Sun, M. Zeilik and D.M.Gibson eds., Springer-Verlag, Berlin, p. 475.
Rodonò M.: 1986c, in Flare Stars and Related Objects, L.V. Mirzoyan (ed.), Armanian Ac. Sci., Yerevan, p. 19.
Rodonò M., Pazzani V., Cutispoto G.: 1983, in IAU Coll. 71, P.B.Byrne and M.Rodonò eds., Reidel Publ. Co., Dordrecht, p. 179.
Rodonò M., Cutispoto G., Catalano S., Linsky J.L., Gibson D.M., Brown A., Haisch B.M., Butler C.J., Byrne P.B., Andrews A.D., Doyle J.G., Gary D.E., Henry G.W., Russo G., Vittone A., Scaltriti F., Foinq B.: 1984, Proc. 4th Eur. IUE Conf., ESA SP-218, 247.
Rodonò M., Cutispoto G., Pazzani V., Catalano S., Byrne P.B., Doyle J.G., Butler C.J., Andrews A.D., Blanco C., Marilli E., Linsky J.L., Scaltriti F., Busso M., Cellino A., Hopkins J.L., Okazaki A., Hayashi S.S., Zeilik M., Henson G., Smith P., Simon T.: 1986, **Astron. Astrophys. 165**, 135.
Rodonò M., Byrne P.B., Neff J.E., Linsky J.L., Simon T., Butler C.J., Catalano S., Cutispoto G., Doyle J.G., Andrews A.D., Gibson D.M.: 1987, **Astron. Astrophys. 176**, 267.
Saar S.H., Linsky J.L.: 1987, Colorado Astrophysics Preprint.
Saar S.H., Linsky J.L., Beckers J.M.: 1985, Colorado Astrophysics Preprint.
Schrijver C.J.: 1986, **Stellar Magnetic Activity**, Ph.D. Thesis, Utrecht University.
Schussler M.: 1983, IAU Symp. 102, J.O.Stenflo ed., Reidel Publ. Co., Dordrecht, p. 213.
Simon T.: 1986, Joint NASA-ESA-SERC Conf., London, **ESA SP-263**, 53.
Stauffer J.R., Hartmann L.W.: 1987, preprint.
Stauffer J.R., Schild R.A., Baliunas S.L., Africano J.: 1987, preprint.

Stenflo J.O. (ed.): 1983, Solar and Stellar Magnetic Fields: Origin and Coronal Effects, IAU Symp. 102, Reidel Publ. Co., Dordrecht.

Vaiana G.S., Cassinelli J.P., Fabbiano G., Giacconi R., Golub L., Gorenstein P. Haisch B.M., Hardner F.R., Johnson H.M.Jr., Linsky J.L., Maxon C.W., Mewe R., Rosner R., Seward F., Topka K., Zwaan C.: 1981, Astrophys. J. 245, 163.

Vogt S.S.: 1983, in IAU Coll. 71, P.B.Byrne and M.Rodonò eds., Reidel Publ. Co., Dordrecht, p. 137.

Wilson O.C.: 1978, Astrophys. J. 226, 379.

Zeilik M., Gibson D.M. (eds.): 1986, Cool Stars, Stellar Systems, and the Sun, Springer-Verlag, Berlin.

SOLAR AND STELLAR CONVECTION

J.-P. Zahn
Observatoires du Pic-du-Midi et de Toulouse
14, avenue E. Belin   31400 Toulouse, France

ABSTRACT. We examine the progress made during the last decade in our knowledge of stellar convection, first in the theory and using numerical simulation, and second through observations, from ground and space.

Special emphasis is put on those results which allow to tackle the basic questions: i. Is the same mechanism responsible for the different types and scales of motions that are currently identified in the solar convection zone: granulation, mesogranulation, supergranulation, large eddies? ii. On which scale(s) is energy injected into the convective motions? For what reason?

1. INTRODUCTION

Eleven years (a solar cycle!) have elapsed since the Nice meeting on stellar convection, and it would be worthwhile to give a complete account of the progress which has since been achieved in this field. This however would require a whole colloquium, and I have therefore chosen to focus this review on what may be called 'basic convection'. In doing this, I shall follow the same approach as E. Spiegel during the historical joint discussion (on aerodynamic phenomena in stellar atmospheres) held in 1964 at the IAU assembly in Hamburg; it is diificult to forget how much we owe to him for promoting the hydrodynamical treatment of stellar convection.

Let me quote him, since he formulated exactly what I had in mind, before re-reading his paper (Spiegel 1966): "In spite of (the) very real separation between the realms of observation and theoretical knowledge, a discussion of the interaction between the convection zone and the photosphere may be useful at this time, if only to illustrate the present possibilities for theoretical interpretation. To proceed usefully, it is necessary to limit severely the range of solar phenomena to be discussed and to include only those which seem to be direct manifestations of convective motion and induced phenomena."

I shall apply this rule very strictly, therefore excluding from this presentation many important topics which are closely related to convection, but do not seem to contribute much to the understanding of its basic mechanism. Such are the penetration into the stable radiative interior, and also the generation of acoustic waves, of differential rotation and of magnetic fields, together with their feedback on convective motions; I shall of course ignore any kind of activity. The basic convection that I shall consider is what may occur in a hypothetical non-rotating, non-magnetic star. For those who may feel frustrated, quite understandably, by such a narrow approach, I recommend the excellent review by Å. Nordlung (1985a), which covers the subject in a more extensive way (see also Mattig 1985, Schröter 1985, Nordlung 1985b and Unno 1986).

2. THEORY

During the past decade, progress has been slower on the theoretical as opposed to the observational side. True, we now have a much better understanding of the behavior of dynamical systems, and we begin to perceive the role of strange attractors in turbulence. It is no coincidence that the first strange attractor has been identified in a system of equations meant to represent Boussinesq convection (Lorenz 1963). The subject is expanding at a fast rate, and a whole colloquium has been recently devoted to chaos in astrophysics (eds. Buchler, Perdang and Spiegel 1985).

However, at the present stage, these powerful concepts are still difficult to use in stellar convection for the immediate purpose of comparing theoretical predictions with the observations. In the meanwhile, we must still proceed along the classical routes: improving the mixing length theory, or performing numerical simulations.

2.1 Mixing length theory

The mixing length procedure is still much in favor in stellar evolution theory, since it provides a means of building stellar models at little cost. Its shortcomings, at least in the original formulation by E. Böhm-Vitense (1958), are well known, and they have fostered various attempts to remedy them. Such improvements however cannot cure the main weakness of that approach, namely the rather arbitrary definition of the mixing length.

Nonlocal mixing lengths are introduced mainly for the purpose of describing the overshooting of convective motions into the stable neighbouring layers. Following Unno (1969) and Maeder (1975), new formulations have been proposed by Xiong (1979), Eggleton (1983) and Kuhfuß (1986), which also enable the description of time dependent behavior. Another approach by Roxburgh (1978) has been correctly criticized by Eggleton (1983) and by Baker and Kuhfuß (1987).

A convenient way to take into account the Reynolds stresses due to the small scale motions is to implement a turbulent pressure of order ρv^2, ρ being the density and v a characteristic velocity. The effects of such a turbulent pressure have been analyzed by Antia, Chitre and Narasimha (1983). They find that it alters enough the superadiabatic gradient to significantly increase the growth rate of the linear fundamental mode, for scales which are of the order of the supergranulation. As shown by Narashima and Antia (1982), it has furthermore the effect of smoothing the eigenfunctions near the surface, thus accelerating the convergence of the superposition of linear modes to reproduce the convective flux, a convergence which had been questioned before (Hart 1973).

It is worth mentioning that, to some extent, mixing length approximations are used also in most numerical simulations discussed hereafter. As we shall see, only the largest scales are represented explicitly in such calculations, and the contribution of the smaller ones is often rendered by some type of eddy diffusion, whose value is deduced from a typical (mixing) length and a typical velocity, much as in the classical mixing length theory.

2.2 Direct simulations

The ambition of direct simulations is to solve the hydrodynamical equations with the least possible approximations. To take up this challenge without any concession, it would require a computer powerful enough, in speed and memory, to encompass all scales which are present in stellar convection, from the largest which are of the size of the star, to the smallest at which the kinetic energy is dissipated through viscous friction (about a centimeter in the Sun!).

Such a computer is not available yet (will it ever be?). One must therefore choose the scales to be explicitly described, and one needs a prescription to take into account the smaller, subgrid scales. Several recipes now exist, with their pros and their cons: they will not be discussed here. Ideally, one should restrict their use to those scales for which some similarity law is firmly established (such as a turbulent cascade); even this is very rarely possible with the currently available computers.

2.2.1 Modal expansions

The modal approach consists in representing the horizontal pattern of the convective motions, which is assumed to be periodic in space, by a severely limited number of planforms (one or two Fourier components in the simplest case), but to seek a much higher resolution in the vertical direction (typically 200 mesh points, or more). Most often, such simulations are accelerated by using the anelastic approximation (Gough 1969): it filters out the sound waves, and thus only keeps the (slower) gravity modes.

This method has been applied to various situations, both to ideal polytropes and to realistic stellar models; let us quote the salient results which have been obtained.

i. Near the top of the unstable domain, the pressure fluctuations, which are required to deflect the vertical upwellings into horizontal streams, are responsible for density fluctuations that may be large enough to change the sign of the buoyancy force. This 'buoyancy braking' is due clearly to the stratification; it vanishes in the incompressible (Boussinesq) limit (Massaguer and Zahn 1980; Massaguer *et al.* 1984). The pressure fluctuations, which are ignored in the mixing length theory, are of such importance that the net work done by them is in some cases larger than the net work done by the buoyancy.

ii. In a strongly stratified medium, the convective cells may extend over several pressure scale heights, and strong horizontal shearing flows appear at their top to ensure the conservation of mass. To illustrate this, the horizontal velocity associated with scales which are of the size of the solar convection zone reach 100 m/s at a depth of 7 Mm, according to a single-mode calculation made by Latour, Toomre and Zahn (1983).

iii. The convective motions penetrate rather deep into the adjacent stable layers, and according to such calculations the extent of penetration is a function of both the thickness of the unstable domain and of its stratification. It is found to depend sensitively on the geometry of the considered cellular flow: cells with a central upwelling penetrate much deeper than those with a central downstream, as confirmed by the supergranular velocity field (Toomre *et al.* 1976; Zahn *et al.* 1982; Massaguer *et al.* 1984).

The modal procedure is unquestionably an improvement with respect to the mixing length treatment. However, when reduced to only one or two planforms, as it has usually been the case due to limited computer power, the modal calculations suffer from the fact that the horizontal size of the cells, and also their geometry, is arbitrarily imposed. (A notable exception is the calculation performed by Marcus (1980) with 12 complete spherical harmonics.) One consequence of such a severe truncation is that the smallest scales, at which kinetic energy is dissipated through viscous friction, have a very large aspect ratio, leading to artificial horizontal boundary layers when the viscosity becomes too small. Another is that most solutions are stationary at Rayleigh numbers where they should be time dependent; the time dependence can be recovered when including modes of vertical vorticity, as shown recently by Massaguer and Mercader (1984).

However, in spite of all these shortcomings, the modal method has enabled the rather important results presented above to be established, and those have been confirmed since by the much more realistic calculations to be discussed next.

2.2.2 Two-dimensional simulations

Such simulations also assume a periodic pattern in the horizontal direction, but they do not impose the geometry of the flow, as in the modal procedure. Both finite differences and spectral methods are used to discretize the equations; the latter are particularly well adapted to the present vector computers. Run with such powerful computers, the two-dimensional codes enable one to explore the full effects of compressibility, without being restricted to the anelastic approximation.

The main results obtained by performing such calculations may be summarized as follows:

i. The flow always remains subsonic, except when imposing an unrealistically steep superadiabatic gradient throughout the domain. This property can be considered as a justification of the use of the anelastic approximation, which is only valid at small Mach number (Graham 1975; Hurlburt, Toomre and Massaguer 1986; Yamagushi 1984, 1985).

ii. The solutions all display a time-dependent behavior at Rayleigh numbers larger than about hundred times critical; they are characterized by strong, narrow, unstable streams, which are directed downwards and penetrate deeply into the stable region below (Hurlburt et al. 1986).

The great advantage of two-dimensional calculations is that they can be performed with a high enough spatial resolution to enable, at least in principle, to reach the viscous dissipation scale. Unfortunately, they are untypical of real, three-dimensional flows, such as highly nonlinear convection, since in the inviscid limit, they not only conserve energy, but also enstrophy (the square of the vorticity). This completely alters the coupling between the scales; it produces an inverse cascade of energy from the small to the large scales (which in the simulations can be observed as the pairing of vortices), and a steep enstrophy cascade to the smallest scales.

In spite of this serious bias, two-dimensional calculations serve as a powerful first exploration of a given problem, provided that energy is injected at the largest scales, and that the Reynolds number is kept reasonably small (in order to avoid the cascades mentioned above).

2.2.3 Three-dimensional simulations

Those are of course the most desirable simulations to perform; the only limitations are their low spatial resolution (32x32x32 for a Cray 1) and their cost. In this field again, E. Graham (1977) was the pioneer: he clearly demonstrated the truly three-dimensional nature of convection, and the important role of the vertical vorticity in the nonlinear coupling of the various modes.

The Sun has been the main target of such calculations. Å. Nordlung (1982) was the first to simulate the granular motions, and he took great care in treating the radiative transfer of energy. He made a fairly good contact with the observations: exploding granules, sharp intergranular lanes, life times, etc. (Wöhl and Nordlund 1985). The major discrepancy is the large temperature contrast needed to carry the convective flux, which is higher than that allowed by the observations, even after correction from instrumental diffusion and poor seeing. The calculated vertical velocities (Nordlund 1985b) also seem larger than the observed ones, but no thorough comparison has been made yet, probably because of the lack of resolution of the observational data.

More recently, Chan and Sofia (1986) performed similar calculations, and they found that the vertical correlation of the motions extends over a little more than a pressure scale height; they consider this as a justification for the usual procedure of linking the mixing length to the pressure scale height. It remains to be seen, of course, whether the ratio between those two scales is indeed independent of the depth, and of the stratification.

The simulations made by Glatzmaier and Gilman (1982), Glatzmaier (1984, 1985) and by Gilman and Miller (1986) take the global scale as their computational domain. Since this scale contributes little to the convective energy transport, an eddy diffusivity of heat is invoked to lessen the superadiabatic stratification; it is assumed to represent the effect of the small scales (supergranulation, granulation). With such a large eddy diffusivity, the effective Rayleigh number is quite low (50 times critical), and the solutions display a fairly regular cellular pattern, with long cells aligned along the rotation axis (the so-called banana cells).

This geometry was already predicted quite a while back by Busse (1970) and by Durney (1970), who found that those modes are the most unstable and would dominate in the nonlinear regime. Moreover, this trend has been confirmed in a cleverly designed microgravity experiment, which was performed in Spacelab 3. There, a spherical shell of fluid was heated from inside and was submitted to a radial electrostatic force; when rotation was imposed, the degeneracy was lifted again in the way predicted by Busse (Hart *et al.* 1986; Toomre, Hart and Glatzmaier 1987).

3. OBSERVATIONS

In the last five years, significant progress has been achieved in observing stellar convection. Understandably, most results concern the Sun, whose proximity permits both extremely good spectral resolution and very high spatial definition. In the stellar case, information must still be extracted out of the scrambled signal originating from the whole disk of the star, and much less photons are available.

But thorough analysis of line shifts and line asymmetries is providing more and more clues about stellar photospheric motions and their vertical distribution. This is explained in detail during this meeting by D. Dravins, who with D. Gray is one of the pionneers in the field (Gray 1980, 1981, 1982, 1986; Dravins, Lindegren and Nordlung 1981; Dravins 1982; Gray and Toner 1986; Dravins, Larsson and Nordlung 1986).

And it appears that global pressure oscillations have been detected by Gelly, Grec and Fossat (1986) on two bright stars, Procyon and α Centauri, much in the same way as they were on the Sun. The road is thus opening for using yet another technique of probing the stellar interiors, and to glean new information about the convection zones.

Unquestionably, it is this technique which has given a new and strong impetus to solar research. Heliosimolgy has enabled us to explore the vertical stratification of the convection zone, and for the first time to locate its bottom through a direct method. Its rotation state can also be investigated now, but the signal to noise ratio is still insufficient to establish a firm result. A whole colloquium was recently devoted to this exciting research, and therefore I shall not expand on it (Advances in Helio-Asterosismology, Aarhus 1986, proceedings in press).

Observation from space, free from atmospheric effects, is the other modern technique which has contributed enormously to solar research. The last magnificent result bearing directly on convection is that obtained on Spacelab 2 (Title *et al.* 1986). The photospheric motions have been filmed with the SOUP instrument, and the pictures have been processed by a very sophisticated method. The five-minute oscillations have been subtracted, to leave just the granular motions, which display a large scale organization identical to the chromospheric network, i.e. the supergranulation. The existence of a turbulent cascade is now firmly established for the granules, with the large eddies breaking into smaller ones; the evidence is striking just by looking at the film, and playing it back and forth. A. Title has been invited to this meeting, and he will report in more detail on this breathtaking observation.

But the more classical solar observers have also been extremely active, and at the risk of being blamed for chauvinism, I shall focus the rest of my review on the following results, which I also consider to be very important for the understanding of stellar convection:

i. The distribution of granule sizes and shapes has been determined down to the resolution limit of the solar refractor at Pic-du-Midi.

ii. The vertical velocity field associated with the granular motions has been measured with nearly the same definition, unveiling its kinetic energy spectrum.

iii. A poleward migration of large scale magnetic patterns has been detected by close inspection of the Meudon spectroheliograms.

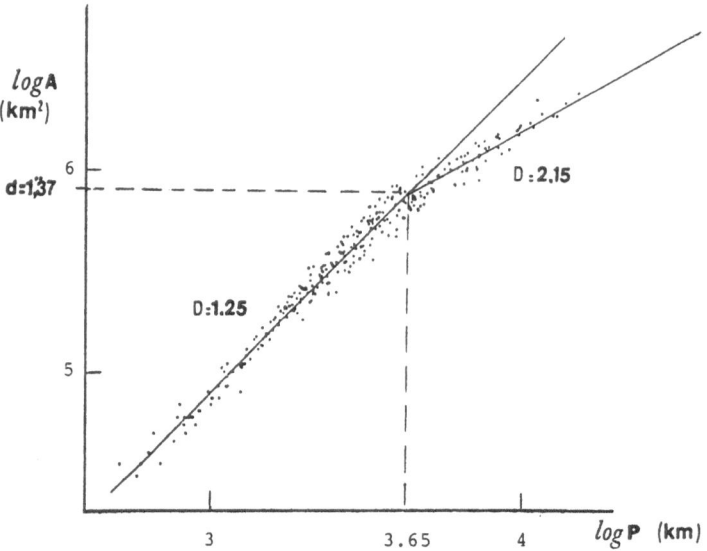

Figure 1. Area-perimeter relation of the solar granulation. Each point represents a granule measurement; D is the fractal dimension (Roudier and Muller 1986).

3.1 Granulation

In the past ten years, the solar refractor of the Pic-du-Midi has yielded images with the nominal resolution of its 50 centimeter objective, i.e. 0.25 second of arc. Thorough analysis of the best images recorded in the past years has revealed that granulation consists of a continuum of sizes, with no dominant scale, contrary to previous claims based on images of poorer resolution. And this distribution has been shown by Macris and Rösch (1984) to vary with the solar cycle: the granules are somewhat smaller at solar maximum than at the minimum.

To render the analysis more quantitative, an attempt has been made, on E. Spiegel's suggestion, to characterize the form of the granules by their fractal dimension. This parameter can be readily deduced from a logarithmic graph of the area versus the perimeter of the granules; the result is shown on figure 1 (Muller and Roudier 1985; Roudier and Muller 1986). It confirms the visual impression that the small granules are much more regular than the large ones, but the distinction between the two is surprisingly sharp with this diagnostic tool. The transition occurs for a size of 1.4 second of arc ($l \approx 10^8\,cm$, or $1\,Mm$). That scale corresponds to a Péclet number vl/K of order 1, if one evaluates it with the vertical velocity ($v \approx 1\,km/s$); at optical depth $\tau = 1$, the heat diffusivity $K \approx 10^{13}$ in cgs units.

This dimensionless number is used to evaluate the strength of the coupling between the velocity field and the temperature field in a conducting fluid: when this number is larger than unity, heat is advected by the motions; when it is smaller, diffusion dominates. Therefore the interpretation of the observed discontinuity in the fractal dimension is that the large eddies are irregular for they advect the temperature field, whereas the small ones are regular because they let the heat diffuse more easily.

Not much more can be deduced from the images alone, but the solar refractor has recently been equipped with the MDPS (Multichannel Double Pass Subtractive) instrument built by the Meudon observatory. The first observation campaign was extremely successful: the best Doppler images obtained with this instrument, on photographic film, have a definition of about half a second, and the vertical velocity field can now be determined with that high resolution (Muller *et al.* 1987).

As is well known, this velocity field is made of two components: the waves due to the five minute oscillations, and the granular motions in which we are interested here. Ideally, to disantangle the two fields, the pictures ought to be processed in the same way as the SOUP images; instead, a much cruder method has been used so far to filter out the waves: it consists simply in superposing two Doppler images taken two and half minutes apart (after careful recentering). When examining the kinetic energy spectra derived before and after this filtering, one notices that it principally affects the supergranular and mésogranular scales, as expected, and that it barely modifies the small scale region. This is fortunate, since it is in that region of the spectrum that the most interesting property has been found.

The result is shown in figure 2, where $E(k)$ is plotted as usual versus the wavenumber k in logarithmic coordinates, $E(k)dk$ being the kinetic energy present in the scales between k and $k + dk$. Two straight lines can be drawn through the data points: for the scales larger than 3 seconds of arc ($\approx 2\,Mm$), the slope is about -0.70, whereas below that scale, and down to the resolution limit, the slope turns out as -1.70.

Within the error bars of the crude treatment which has been used, this value is undistinguishable from the famous $-5/3$ of the Kolmogorov-Obukhov law. This slope characterizes the energy spectrum of homogeneous isotropic turbulence in the so-called inertial domain, where the kinetic energy cascades down to the smallest scales, to be dissipated there through viscous friction (see for instance Landau and Lifschitz 1953, or Tennekes and Lumley 1972). Therefore the observations above must be interpreted as a strong indication that the granular motions are of turbulent origin. They do advect heat in their vertical displacement, as revealed by the well known correlation between temperature and velocity, but according to this result the buoyancy force plays only a minor role in their dynamics, and most of their energy originates from scales larger than $2\,Mm$.

Let us stress that this is the first time that such a slope has been

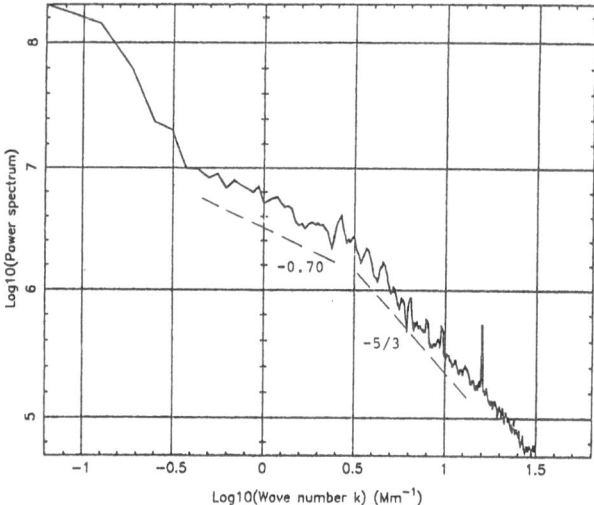

Figure 2. Power spectrum of the solar photospheric motions, derived from two super-posed Doppler images taken two and half minutes apart (Muller *et al.* (1987).

reported. Earlier work by Aime, Ricort and Harvey (1978), with a different technique (speckle-interferometry through photoelectric scans), revealed rather an exponential tail of the kinetic energy spectrum. And recent observations by Nesis *et al.* (1987), likewise recorded on a photographic film but apparently carried out with a somewhat poorer seeing, also show a steeper slope.

3.2 Large scale motions

Several attempts have been made in the last few years to detect a large scale pattern in the photospheric motions, which could be related to the global convection of the Sun and its dynamo; the subject has been recently reviewed by Schröter (1985).

The first direct, kinematic evidence for such motions was the modulation of the differential rotation discovered by Howard and LaBonte (1980) on the Mount Wilson full-disk velocity data. They interpret it as a torsional wave synchronous with the 11 year cycle, which travels from pole to equator in about twice that time.

The second discovery was made by Ribes, Mein and Mangeney (1985) in analyzing the Meudon spectroheliograms, which are now being digitalized. They find that the motions of the newly-born sunspots are organized in several latitudinal bands, as if they were the tracers of a meridional circulation with several counter-rotating, azimuthal rolls (3 or 4 in each hemisphere). The velocities can reach $100\,m/s$; their average is about $30\,m/s$, leading to a turn-over time, for the rolls, of the order of one month.

The authors also find that the mean rotation rate of these young spots is nearly constant with latitude, and thus does not reflect the differential rotation observed in the photosphere. Moreover, they show that this meridional circulation is closely related to the large scale, long-lived magnetic structures outlined by the Hα filaments: these structures coincide with the regions of convergence and divergence (and hence of downflow or upflow) of the azimuthal rolls. Finally, the authors claim that this roll system migrates towards the pole in each hemisphere, with a new pair of rolls emerging at the equator every solar cycle.

This last property seems to contradict that of the torsional wave of Howard and LaBonte, which moves in the opposite direction, from pole to equator. Moreover, those authors have found no indication of meridional flows with velocities larger than $10\,m/s$ (their precision limit). But the tracers used in the two cases are not the same: the Doppler velocities are determined in the photospheric plasma, whereas the young spots may be anchored very deep in the convection zone.

It has also been noticed, of course, that the rolls described by Ribes *et al.* are parallel to the equator, whereas convection in a rotating shell should proceed in cells aligned with the rotation axis (cf. section 2.2.3 above). But this discrepancy can be explained. For instance, one may argue that the drift of the magnetic pattern, as visualized by the young sunspots, is not necessarily connected with large scale convective motions. A more plausible explanation is that the toroidal (azimuthal) magnetic field, which in the classical dynamo theory is generated by the differential rotation, may be strong enough to lift the degeneracy due to the spherical symmetry in a way opposite to that of the Coriolis force, namely in azimuthal rolls. Further work is necessary to clarify this point.

Other magnetic tracers have been used in the past with no definite conclusion (see Schröter 1985). During this meeting, P. Ambrož reports on a circulation pattern which he derives from the Hα Synoptic Charts, using a somewhat subjective method. His conclusion is that the displacements of these structures cannot be interpreted by purely zonal flows, and the meridional velocities he finds can reach $150\,m/s$ (Ambrož 1986). A detailed comparison with the results of Ribes *et al.* (1985) is not possible, since they concern different epochs.

4. DISCUSSION

What can be learned, about the nature and the mechanism of solar convection, from the observations that have been presented above?

The result obtained by Muller *et al.* (1987), which apparently proves the turbulent nature of the granular motions, is of such importance that the whole chain of treatment of the Doppler images must be firmly validated before going too far into the interpretation. The authors themselves insist that their result be considered as very preliminary, and they wish to perform various tests before declaring it definite. But let us take it here at face value, to confront it with other evidence.

The fact that the granules participate in a turbulent cascade can be clearly seen in the SOUP film mentioned above, after filtering out the waves: the large eddies always split into smaller ones, and the opposite (the merging of small granules) never occurs. What is more surprising is the almost perfect $-5/3$ slope, which is commonly regarded as a signature of homogeneous isotropic turbulence; such ideal turbulence is not likely to exist in this strongly stratified medium, where the density scale height is about $0.5\,Mm$, i.e. less than the horizontal size of the observed eddies.

But one may argue that the assumption of isotropy is not necessary to establish the Kolmogorov-Obukhov law: it suffices that the anisotropy be independent of scale. And homogeneity is clearly achieved in the horizontal directions, which are those where most dynamical interactions occur in the photosphere: the vertical motions are hindered by the convectively stable region located above optical depth $\tau = 1$, and by the 'buoyancy braking' described by Massaguer and Zahn (1980).

At first glance, this result of a turbulent cascade, which implies that the kinetic energy of the granules proceeds from larger scales, is in contradiction with theoretical work. The linear analysis made by Antia, Chitre and Narasimha (1983), which includes the effect of the turbulent pressure, shows that the growth rate of the fundamental mode is maximum for the granular scale, which seems to indicate that those scales should be the dominant ones in the nonlinear regime. However, when evaluating the turbulent heat conductivity and the turbulent viscosity due to the smaller scales, they neglected the contribution from eddies larger than the classical mixing length (i.e. the pressure scale height). That this truncation is not justified can be seen in the observation of the granular shapes reported above, from which one can derive, as explained, the Péclet number vl/K; the mixing length deduced from this number is comparable with the horizontal size of the eddies, and it is thus significantly larger than the pressure scale height (which is about $150\,km$ in the photosphere). Let us notice also that the observed lifetime of the granules (around $10\,mn$) is consistent with the turn-over rate estimated with this horizontal scale, $v/l \approx 10^{-3}s^{-1}$, with again $v \approx 1\,km/s$. The maximum linear growth rate predicted by Antia *et al.*, with the truncated eddy diffusivities, is just of that order; one may therefore conjecture that the granular scales would turn out as stable (damped) when using the correct value of those coefficients.

The numerical simulations made by Nordlung (1982) also prove that the granules can live by themselves, from the work done by the buoyancy and pressure forces alone, and that they need no other source of energy. But this result is obtained by *imposing* that the convective heat flux be carried only through the eddies of granular size, and therefore it is not too surprising. As already mentioned, Nordlung predicts a temperature contrast, and also vertical velocities, which are larger than those allowed by the observations; this may be taken as an indication that other scales, presumably larger ones, participate in the convective transport. If these scales were included in the simulation, they would probably lower the superadiabatic gradient, and hence the temperature fluctuations and the velocities associated with the granular motions.

If we accept the idea that the granules are not the major actors in the solar convection, we must identify the scales which are the dominant ones, namely those at which the major part of the kinetic energy is injected into the turbulence. In all likelihood, these are the supergranuler eddies, which are known to be convectively unstable (Van de Borght 1979; Antia, Chitre, and Pandey 1981; Antia, Chitre and Narasimha 1983).

To confirm this property, it would suffice to check that the energy spectrum is indeed maximum for the corresponding wavenumbers. Unfortunately, what seems the best spectrum obtained so far, that of Muller *et al.* (1987), is not reliable enough in that wavenumber range, mainly because of the rough method used to filter out the five minute oscillations. Nevertheless, we can use the best established part of this spectrum, namely the inertial range, and compare it with the energy of the supergranular motions, using the information which is available on sizes and velocities.

We shall proceed as follows. First we schematically represent the spectrum of Muller *et al.* by two straight lines (figure 3); the knee $(k_G, E(k_G))$, where the slope steepens from -0.70 to $-5/3$, is located at $2\,Mm$, hence $2\pi/k_G = \lambda_G = 2\,Mm$. There

$$\frac{1}{2}V_G^2 = \int_{k_G}^{\infty} E(k)dk = \frac{3}{2}k_G E(k_G),$$

V_G being the r.m.s. velocity characterizing the granular motions. We take for V_G the value $1\,km/s$ which may be inferred from the vertical velocity measurements (see for instance Muller 1985); we neglect the contribution of scales larger than λ_G.

Likewise, for the supergranules we write

$$\frac{1}{2}V_S^2 = \int_{k_S+\Delta k}^{k_S-\Delta k} E(k)dk = \eta_S k_S E(k_S),$$

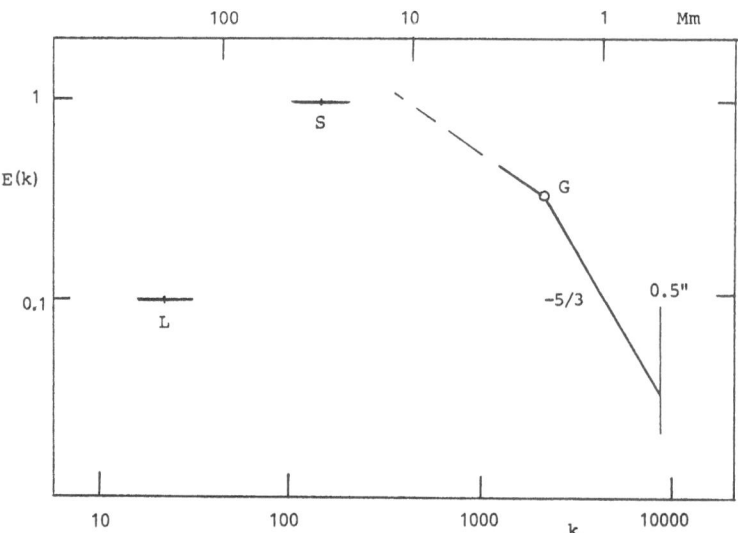

Figure 3. Power spectrum of the solar convection, showing the portion determined by Muller *et al.* (1987), the supergranular motions, and the large scale motions described by Ribes *et al.* (1985).

where V_S is the typical velocity $(250m/s)$ deduced from the lifetime of these eddies (1-2 days), and k_S the typical wavenumber $(2\pi/k_S = \lambda_S \approx 30\,Mm)$ of the supergranules; the integral is taken over the wavenumber range $2\Delta k$ which characterizes them. The coefficient η_S is proportional to the bandwith $\Delta k/k$, and it also depends somewhat on the slope of the energy spectrum in that domain; a reasonable assumption is to take $\eta_S \approx 0.5$.

We thus obtain the ratio

$$\frac{E(k_G)}{E(k_S)} = \frac{2\eta_S}{3}\frac{\lambda_G}{\lambda_S}\left(\frac{V_G}{V_S}\right)^2 \approx 0.35\,.$$

The result is shown on figure 3: $E(k_S)$ is definitely located above the inertial range, and this strengthens our claim that the supergranular eddies are indeed the dominant ones in solar convection, at least near the surface.

It is tempting to use the same method to plot on this graph the large scale motions reported by Ribes *et al.* (1985); we shall use the following parameters: $V_L = 30m/s$ and $2\pi/k_L = \lambda_L \approx 200\,Mm$, and again $\eta_L \approx 0.5$. Here we get

$$\frac{E(k_L)}{E(k_S)} = \frac{\eta_S}{\eta_L}\frac{\lambda_L}{\lambda_S}\left(\frac{V_L}{V_S}\right)^2 \approx 0.10\,,$$

and we see that the energy level of these motions is also located below the maximum reached for the supergranular scales, which seems to indicate that these large eddies are likewise fed by the supergranular motions. But this result must be interpreted with care, since the turbulent convection is certainly far from homogeneous: at the depth where those large scale motions occur, they may well be the dominant ones, and their main energy source may again be the buoyancy.

To summarize: the picture of the solar convection zone which emerges from the most recent observations is that of a classical turbulent fluid, with energy injected at the supergranular scales and cascading down to the smallest scales. It is not yet clear whether the large scales are also fed by those supergranular motions, or are unstable enough to be driven by the buoyancy.

5. PERSPECTIVES

In conclusion, among all the efforts which must be strongly encouraged for a better understanding of stellar convection, I wish to insist on what I consider to be the top priorities. Here they are, in increasing difficulty:

i. With the new parallel computers, one can now perform numerical simulations in a 64x64x64 mesh. Rather than just increasing the resolution of the present calculations, one should verify the possibility of feeding the convection from the supergranular scales. In practice, a mesh spanning the scales from 30 to 0.5 Mm should suffice for a first exploration.

ii. Using the best sites available, and also in planning new space-borne experiments, one should try everything to improve the accuracy and the spatial resolution of the photospheric velocity and intensity measurements.

iii. Asterosismology is an extremely promising diagnostic tool to probe stellar convection zones. Research should be strongly developed in this field, both from the ground (Doppler measurements with the largest telescopes), and out of the atmosphere (photometry).

REFERENCES

Aime, C., Ricort, G., Harvey, J.: 1978 *Astrophys. J.* **221**, 362

Ambrož, P.: 1987, *Bull. Astron. Inst. Czechosl.* **38**, 1

Antia, H.M., Chitre, S.M., Pandey, S. D.: 1981, *Solar Phys.* **70**, 67

Antia, H.M., Chitre, S.M., Narasimha, D.: 1983, *Mon. Not. R. Astr. Soc.* **204**, 865

Baker, N.H., Kuhfuß, R.: 1987, *Astron. Astrophys.* (in press)

Böhm-Vitense, E.: 1958, *Z. Astrophys.* **46**, 108

Buchler, J.R., Perdang, J.M., Spiegel, E.A. (eds.): 1985, 'Chaos in Astrophysics' (Reidel Publ. Co.), *NATO ASI Series C* **161**

Busse, F.H.: 1970, *Astrophysical J.* **159**, 629

Chan, K.L., Sofia, S.: 1986, *Astrophys. J.* **307**, 222

Dravins, D.: 1982, *Ann. Rev. Astron. Astrophys.* **20**, 61

Dravins, D., Lindegren, L., Nordlung, Å.: 1981, *Astron. Astrophys.* **96**, 345

Dravins, D., Larsson, B., Nordlung, Å.: 1986, *Astron. Astrophys.* **188**, 83

Durney, B.R.: 1970, *Astrophys. J.* **161**, 1115

Eggleton P.P.: 1983, *Mon. Not. R. Astr. Soc.* **204**, 449

Gelly, B., Grec, G., Fossat, E.: 1986, *Astron. Astrophys.* **164**, 383

Gilman, P.A., Miller, J.: 1986, *Astrophys. J. Suppl.* **61**, 585

Glatzmaier, G.A.: 1984, *J. Comput. Phys.* **55**, 461

Glatzmaier, G.A.: 1985, *Astrophys. J.* **291**, 300

Glatzmaier, G.A., Gilman, P.A.: 1982, *Astrophys. J.* **256**, 316

Gough, D.O.: 1969, *J. Atmos. Sci.* **26**, 448

Graham, E.: 1975, *J. Fluid Mech.* **70**, 689

Graham, E.: 1977, in E.A. Spiegel and J.-P. Zahn (eds.), 'Problems of Stellar Convection' (Springer), *Lecture Notes in Physics* **161**, 151

Gray, D.F.: 1980, *Astrophys. J.* **235**, 508

Gray, D.F.: 1981, *Astrophys. J.* **251**, 583

Gray, D.F.: 1982, *Astrophys. J.* **255**, 200

Gray, D.F.: 1986, *Publ. Astron. Soc. Pacific* **98**, 319

Gray, D.F., Toner, C.G.: 1986, *Publ. Astron. Soc. Pacific* **98**, 499

Hart, M.H.: 1973, *Astrophys. J.* **184**, 587

Hart, J.E., Toomre, J., Deane, A.E., Hurlburt, N.E., Glatzmaier, G.A., Fichtl, G.H., Leslie, F., Fowlis, W.W., Gilman, P.A.: 1986, *Science* **234**, 1

Howard, R., LaBonte, B.J.: 1980, *Astrophys. J.* **239**, 738

Hurlburt, N.E., Toomre, J., Massaguer, J.M.: 1986, *Astrophys. J.* **311**, 563

Kuhfuß R.: 1986, *Astron. Astrophys.* **160**, 116

Landau, L., Lifschitz, E.: 1953, 'Hydrodynamika' (English translation 1959; Pergamon)

Latour, J., Toomre, J., Zahn, J.-P.: 1983, *Solar Phys.* **82**, 387

Lorenz, E.N.: 1963, *J. Atmos. Sci.* **20**, 130

Macris, C.J., Rösch,: 1984, *C. R. Acad. Sci., série II* **296**, 265

Maeder, A.: 1975, *Astron. Astrophys.* **40**, 303

Marcus, Ph. S.: 1980, *Astrophys. J.* **240**, 203

Massaguer, J.M., Mercader, I: 1984, in J.E. Wesfreid and S. Zaleski (eds.), 'Cellular Structures in Instabilities' (Springer), *Lecture Notes in Physics* **210**, 270

Massaguer, J.M., Zahn, J.-P.: 1980, *Astron. Astrophys.* **87**, 315

Massaguer, J.M., Latour, J., Toomre, J., Zahn, J.-P.: 1984, *Astron. Astrophys.* **140**, 1

Mattig, W.: 1985, in R. Muller (edit.), 'High Resolution in Solar Physics' Springer, *Lecture Notes in Physics* **233**, 141

Muller, R.: 1985, *Solar Phys.* **100**, 237

Muller, R., Roudier, Th.: 1985, in R. Muller (edit.), 'High Resolution in Solar Physics' (Springer), *Lecture Notes in Physics* **233**, 242

Muller, R., Roudier, Th., Malherbe, J.M., Mein, P.: 1987, (poster, this meeting)

Narashima, D., Antia, H.M.: 1982, *Astrophys. J.* **262**, 580

Nesis, A.: 1984, *ESA Spec. Publ.* **220**, 203

Nesis, A., Mattig, W., Fleig, K.H., Wiehr, E.: 1987, *Astron. Astrophys.* (submitted)

Nordlung, Å.: 1982, *Astron. Astrophys.* **107**, 1

Nordlung, Å.: 1985a, *Solar Phys.* **100**, 209

Nordlung, Å.: 1985b, in H. U. Schmidt (edit.), 'Theoretical Problems in Solar Physics' (Max-Planck-Institut für Astrophysik), 1

Ribes, E., Mein, P., Mangeney, A.: 1985, *Nature* **318**, 170

Roudier, Th., Muller, R.: 1986, *Solar Phys.* **107**, 11

Roxburgh, I.W.: 1978, *Astron. Astrophys.* **65**, 281

Schröter, E.H.: 1985, *Solar Phys.* **100**, 141

Spiegel, E.A.: 1966, Transactions IAU **XII B**, 539

Spiegel, E.A., Zahn, J.-P. (eds.): 1977, 'Problems of Stellar Convection' (Springer), *Lecture Notes in Physics* **161**

Tennekes, H., Lumley, J.L.: 1972 'A First Course in Turbulence' (MIT Press)

Title, A.M., Tarbell, T.D., Simon, G.W., and SOUP Team: 1986, *Proc. COSPAR meeting, Toulouse*

Toomre, J., Zahn, J.-P., Latour, J., Spiegel, E.A.: 1976, *Astrophys. J.* **207**, 545

Toomre, J., Hart, J.E., Glatzmaier, G.A.: 1987, in B. Durney and S. Sofia (eds.), 'On the Internal Solar Angular Velocity: Theory and Observations' (Reidel), (in press)

Unno, W.: 1969, *Publ. Astron. Soc. Japan* **21**, 240

Unno W.: 1986, in Y. Osaki (edit.), 'Hydrodynamic and Magnetohydrodynamic Problems in the Sun and Stars' (Univ. Tokyo), 57

Van der Borght, R.: 1979, *New Zealand J. Sci.* **22**, 575

Wöhl, H., Nordlung, Å: 1985 *Solar Phys.* **97**, 213

Xiong, D.-R.: 1979, *Acta Astron. Sinica* **20**, 238

Yamaguchi, Sh.: 1984, *Publ. Astron. Soc. Japan* **36**, 613

Yamaguchi, Sh.: 1985, *Publ. Astron. Soc. Japan* **37**, 735

Zahn, J.-P., Toomre, J., Latour, J.: 1982, *Geophys. Astrophys. Fluid Dynam.* **22**, 159

PHOTOSPHERIC STRUCTURE IN SOLAR-TYPE STARS

Dainis Dravins
Lund Observatory
Box 43, S-22100 Lund, Sweden

The fine structure of stellar photospheres (i.e. the stellar equivalent of solar granulation) can be studied through various observable
parameters. These include photospheric line asymmetries (caused by
different photon contributions from granules and intergranular lanes),
photospheric line wavelength shifts (caused by a bias of photons from
brighter and systematically Doppler-shifted elements), time variability of stellar irradiance (caused by the evolution of a finite number
of granules on the star), and the spatial imaging of stellar surfaces
with optical interferometers. None of these methods is trivial, but it
appears that the least difficult concerns the study of line asymmetries.
These can be observed with reasonable precision and also form a sensitive test for hydrodynamic models of stellar atmospheres (Dravins,
1987a).

Stellar spectra of very high quality are needed in order to see a signature of stellar granulation. A spectral resolution $\lambda/\Delta\lambda \approx 100,000$
may be adequate to reveal the presence of granulation-induced asymme-
tries, but a resolution of 200,000 or 300,000 is preferable in order
to study line asymmetry patterns in different types of lines. Not many
stellar instruments exist in the world with sufficient performance for
such a task.

A number of bright stars have been observed using the ESO coudé echelle
spectrometer in a photoelectric double-pass scanning mode. This mode
allows a resolution $\approx 200,000$, and gives a very clean instrumental
profile, but is also very expensive in observing time. The star most
similar to the Sun among those studied - Alpha Centauri A, G2 V - indeed
reveals a granulation signature rather similar (but not identical) to
that of the Sun. Also other stars observed (of spectral types A, F, G
and K) show line asymmetry patterns indicative of granulation, but with
properties different from the solar ones (Dravins, 1987b).

Numerical supercomputer simulations of stellar surface convection have
been carried out (Nordlund Dravins, in preparation), modeling a small
volume that cuts through the stellar surface: extending up into the

atmosphere, and going down into the convection zone. These time-dependent models of the 3-dimensional radiation-coupled convection contain
only three free physical parameters: the effective temperature (which
determines the heat flow through the bottom of the simulation volume);
the stellar surface gravity (which influences the vertical stratification), and the chemical abundance (which determines the opacities
throughout the volume). These models are stepped forward in time to
reveal the properties of convection cells (granules) on stellar surfaces: their sizes, the patterns in their gas flows, and the structure
of convective motions beneath the visible stellar surface. The time
evolution sequences show the processes of granulation cells breaking
apart and dissolving while new ones are being formed nearby. Granules
on a K dwarf may be only a few hundred km across, while those on a G
subgiant are on the order of 10,000 km, much larger than solar ones.
As the next step, spectral line profiles were computed for each spot
on the stellar surface, and for each timestep in the model sequence,
in effect using the computed simulation as a set of spatially and temporally varying model atmospheres (Dravings Nordlund, in preparation).
By averaging over a very large number of such individual line profiles,
a representative line for various positions on a stellar disk is obtained, as well as a profile for integrated starlight.

Due to the velocity fields present on the star, this integrated starlight profile is appreciably broader than the line from any one spot
on the surface, and due to unequal contributions from hot and cool
areas, the lines are also asymmetric and wavelength-shifted. These
synthetic line profiles are then compared to observations to verify
the degree of realism in the model simulations. The results show a
very satisfactory agreement concerning the line asymmetry patterns,
while some discrepancies in the outer wings of the absorption line
profiles suggest that the horizontal surface motions in some stars
may be even more vigorous than the present anelastic model allows them
to be.

Convective wavelength shifts in these models vary between some 1000 m/s
for an F star model at 6600 K, and some 200 m/s blueshift for a K
dwarf at 5200 K. This change a factor of five in convective lineshift
reflects not only a larger temperature contrast between granules and
intergranular lanes in F stars, but in particular the circumstance
that the "real" granulation in cooler K stars lies slightly beneath the
visible surface, shrouded by a "mist" of light-absorbing gas that only

permits one to faintly perceive the contrasts beneath. In contrast, the granules in an F star appear "naked" and one can look right onto the top of the convective layers. The Sun occupies an intermediate position in this respect, its granulation contrast being somewhat damped by the overlying atmospheric layers.

III. OUTER ATMOSPHERES, WINDS

SOLAR AND STELLAR CHROMOSPHERES*

R. Hammer

Kiepenheuer-Institut für Sonnenphysik

Schöneckstr. 6, D-7800 Freiburg

ABSTRACT

This review attempts to highlight two fundamental and complementary aspects of the
chromospheric phenomenon; viz., global properties of stellar chromospheres and
their variation among the stars, and the underlying fine structure that affects or
determines these global properties. After an introductory discussion of the gross
vertical structure of a stellar chromosphere, attention is given to the chromospheric geometric extent and its dependence on the position of the star in the HR diagram. This includes a critical review of various explanations of the dividing line
that separates red giants from Solar-like stars. A subsequent chapter summarizes
main features of proposed chromospheric heating mechanisms and discusses the role
that magnetic fields play in the transport and dissipation of energy. The onset of
stellar chromospheres and the magnitude of nonthermal motions can be probed by the
Wilson-Bappu effect and by similar empirical results. The well-known insensitivity
of the CaIIK line width to the stellar activity level might be largely due to a
collective effect. The final chapter explores the dependence of chromospheric
magnetic fine structure on global stellar properties.

I. INTRODUCTION

The Solar chromosphere is usually pictured as a layer in which the average temperature T increases slowly outward, until it suddenly jumps to much higher values in
the transition region (hereafter TR) to the corona. Fig. 1 shows the variation of
temperature with height in the chromospheric reference model C of Vernazza et al.
(1981). This model has been constructed to fit the "average" emission (i.e., the
emission at extremely low spatial and temporal resolution) of the quiet Sun under
the simplifying assumptions of a static steady state and plane parallel geometry.
Clearly, these approximations break down in the real Solar chromosphere, which is
highly inhomogeneous, temporally variable, and far from being static. Nevertheless,

* Mitteilungen aus dem Kiepenheuer-Institut Nr. 280

it is important to consider first an average model in order to establish a concep-
tual basis for any further discussion of the various fluctuations in space, time,
and velocity space.

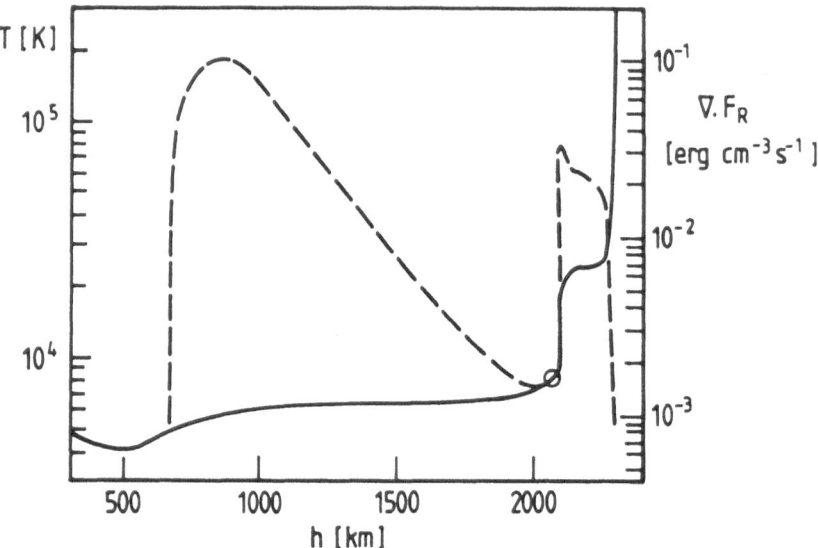

Figure 1. Temperature (drawn) and net radiative energy loss (dashed) vs. height for
the Solar chromospheric reference model C of Vernazza et al. (1981). The circle
marks the point where hydrogen is half ionized. The emission peak at 2200 km is due
to Lα.

In this review I will adopt the current praxis (e.g., Ayres 1979; Linsky 1980;
Athay 1981a; see, however, the discussions in Jordan and Avett 1973) to use the
term "chromosphere" for the region between the temperature minimum and the TR, the
latter being characterized by its steep temperature gradient and the resulting importance of thermal conduction in the energy balance. In extant Solar reference
models (e.g., Fig. 1) the chromosphere extends over a height interval of roughly
2000 km, which is large compared with the pressure scale height, $H_p = |dh/dlog\ T|$,
but small compared with the Solar radius. This differs from the cool low gravity
stars, where the chromospheric extent may exceed the large stellar radius, and
where the hot corona may be missing altogether.

The paper is organized as follows. Sect II deals with the gross vertical structure
of the average chromosphere and with the variation of its geometric extent among
the stars. Sect. III summarizes important aspects of proposed heating mechanisms
and discusses the role that magnetic fields play in the energization of the chromosphere. Sect. IV deals with the Wilson-Bappu effect and related empirical results,
which can be used to probe the depth (in terms of mass column density) of Solar and
stellar chromospheres as well as chromospheric nonthermal velocities. The paper
concludes by combining implications of the preceding sections in a discussion of
the dependence of chromospheric magnetic fine structure on global stellar properties (Sect. V).

II. GROSS VERTICAL STRUCTURE

The chromosphere emits more energy than it absorbs. This energy loss must be sup-
plied continuously by some kind of mechanical heating. In the reference model shown
in Fig. 1, the net radiative energy loss is represented by the dashed curve. Any
proposed heating mechanism should match such an empirical curve at least approximately.

Over most of the chromosphere, the required heat input decreases with height
slightly less rapidly than the density. Such a general decrease suggests the gradual absorption of energy coming from below. This is not necessarily so for all
types of stars. In dMe stars, for example, the coronal activity level is so high
that the absorption of coronal X-rays represents an important heating mechanism in
the upper chromosphere (Cram 1982).

Similarly, the abrupt Lα peak in the radiative loss curve of the Solar reference model (Fig. 1) might also be powered from the corona. Hot coronal electrons
can "leak" through the thin TR (Roussel-Dupré 1980a,b; Shoub 1983; Dufton et al.
1984). Owocki and Canfield (1986) have suggested that this nonlocal thermal conduction can supply the uppermost chromosphere with sufficient energy to balance the
Lα losses. On the other hand, it has also been suggested that upward traveling
waves might suffer heavily from radiation damping as soon as Lα becomes effectively thin (Ulmschneider 1986). And finally, the Lα peak might be a consequence of the breakdown of the assumptions underlying the reference model. In particular, the Lα emitting layer on the Sun is by no means plane parallel (Bonnet
et al. 1980; Foing et al. 1986).

Similar to the Lα emission peak, the behaviour of the net emission curve below
1000 km is also highly uncertain, as has been emphasized by several authors (e.g.,

Athay 1985; see also the discussion and further references in Lites 1985). This uncertainty is due to the large number of opacity sources that become important near
the temperature minimum. As a consequence, the reality of the rapid onset of the
heat input right at the temperature minimum (cf. Fig. 1) is not yet definitely
proven; but it appears to be likely, at least. In principle, a temperature inversion does not necessarily require such a rapid onset of the heat input. An outward
temperature rise can also be produced by a smoothly varying heating law (e.g.,
Ayres 1979) or by nLTE radiation effects alone, without any mechanical heating
(Cayrel 1963). Nevertheless, present Solar reference models do suggest that the
Cayrel effect is rather unimportant (cf. Avrett 1973), that there is indeed a fairly rapid onset of the heat input, and that most of the heating occurs already in
the lower chromosphere, within a few scale heights above the temperature minimum.

The radiative energy loss of the Solar chromosphere occurs mainly in optically
thick emission lines formed in nLTE. Most of these lines are effectively thin;
i.e., the photons escape after multiple scattering. In this case, the local radiative energy loss per volume can be approximated by,

$$|\nabla . F_M| = \nabla . F_R \simeq n_e n_H f(T), \qquad (1)$$

where n_e is the electron density, n_H is the density of hydrogen atoms and protons,
and $f(T)$ is an emissivity function, which at chromospheric temperatures increases
steeply with T.

On the basis of the chromospheric energy balance equation (1) we can try to understand some of the basic properties of the gross vertical structure of a stellar
chromosphere. In particular, Eq. (1) allows us to discuss why the temperature increases outward, why this increase is gentle, why a transition to high coronal temperatures occurs in stars like the Sun, and why this is perhaps not the case in
some other types of stars. In order to keep this discussion simple, let us neglect
any temperature dependence of the heat input. In Solar reference models, the heating is found to decrease outward slightly less rapidly than the density, and thus
much less rapidly than the emission at constant temperature, which decreases like
density squared. Therefore, the temperature must change outward in such a way as to
enhance the emission appropriately.

At cool chromospheric temperatures, this can be achieved easily by a very gentle
outward temperature rise because both n_e and $f(T)$, and thus the emission, depend
very sensitively on T. The sensitivity of n_e is due to the fact that over most of
the chromosphere the electrons are tapped from the big reservoir of hydrogen ionization; and as long as the fractional ionization of H is small, n_e increases very

strongly with T. This "thermostatic effect" (Athay 1981b, 1985) of hydrogen ionization terminates when H becomes mostly ionized, so that $n_e \approx n_H$, independent of T.
Moreover, at sufficiently large temperatures, near $T \approx 10^5 K$, the emissivity $f(T)$
has a maximum. Typical calculations of $f(T)$ (e.g., McWhirter et al. 1975) exhibit
an additional local maximum due to hydrogen emission near $T \approx 15000K$. It is not
yet clear to what extent and on which stars this local maximum can be smoothed out
by optical thickness effects in Lα (cf. McClymont and Canfield 1983; however
Athay 1986).

As soon as the emission reaches its maximum as a function of T, it can no longer be
adjusted to the heat input. Then the chromospheric type of energy balance described
by Eq. (1) breaks down. Even below this point the sensitivity of the emission to
temperature is reduced to such an extent that energy balance can be achieved only
by means of a steep temperature gradient, for which thermal conduction becomes important in the energy equation. This changes the character of the energy balance
from local (Eq. (1)) to global, because thermal conduction collects and redistri-
butes the dissipated energy before it is radiated away.

For any prescribed chromospheric heating law in which the dissipation falls off
less rapidly than density squared, the temperature increases outward, and we must
ultimately reach such a critical position beyond which a chromospheric type of
energy balance is no longer possible. However, this outer limit for the chromosphere need not be the position where the corona begins. In principle, the TR to a
steady corona may be located anywhere below this critical position. The actual
location of the TR is likely determined by the amount of energy that is deposited
in, and lost by, the corona (cf. Hammer et al. 1982, 1983, 1984) because the total
energy loss F_{cor} of the corona is related to its base pressure p_{TR}. It is fairly
obvious that such a relationship must exist: Provided that other conditions are the
same, a corona with a larger base pressure has larger radiation losses from both
its coronal and TR portions; and in open regions the mass flux is also enhanced.
The required relationship between F_{cor} and p_{TR} can be derived from theoretical models of closed coronal loops (e.g., Rosner et al. 1978, Eq. (4.4)) and open regions
(Couturier et al. 1979; Hearn and Vardavas 1981a,b; Hammer 1982a and 1982b Fig. 2)
as well as from semi-empirical models (Jordan 1980 Fig. 4).

Many types of stars have outer atmospheres very similar to the Solar one (cf. review by Linsky 1985a). Along the main sequence, this applies to stars of spectral
type later than early F, perhaps even late A. When we go from the Solar-like stars
to the cool low gravity stars in the upper right corner of the HR diagram, the
basic character of the outer stellar atmosphere appears to change: chromospheres
become geometrically extended; cool, low speed winds set in (Reimers 1977; Stencel

and Mullan 1980); and the UV and X-ray emission indicative of hot coronae disappears (Linsky and Haisch 1979; Ayres et al. 1981; Simon et al. 1982). These changes are observed to occur in the vicinity of a dividing line in the HR diagram, which for giants lies near spectral type KO (for recent reviews see Dupree 1986; Dupree and Reimers 1987; Linsky and Jordan 1987). The neighborhood of the dividing line is populated by hybrid stars (Hartmann et al. 1980; Brown 1986), which show both emission from warm (10^5K) plasma and spectroscopic evidence for cool winds.

Since on the Sun the high temperature X-ray emission comes mostly from closed loops, while the wind comes from magnetically open regions, the observed changes suggest that the character of both types of regions changes in the relevant part of the HR diagram. These changes need neither be simultaneous nor particularly abrupt.

Several of the fundamental stellar properties vary rapidly in this region of the HR diagram. From yellow to red giants, gravity decreases, while the average rotation period, the fractional depth of the convection zone, the fractional size of the convective elements, and the ratio of photospheric pressure scale height to radius increase. It is reasonable to expect that these differences of the basic stellar properties ultimately cause the observed differences of the atmospheric structure; but how they do so is still not understood. Many different suggestions have been put forward in the past; let me briefly discuss some of them.

The most important determinant of stellar activity is the rotation period, which tends to be larger to the right of the dividing line. It has been suggested that this effect alone might account for the low UV and X-ray emission (Zwaan 1986). However, it has also been argued that stellar rotation periods grow already to the left of the dividing line (e.g., Gondoin et al. 1987).

Böhm-Vitense (1986) suggested that in the outer chromospheres of red giants the heating might decrease outward as rapidly as density squared, or even faster. Therefore, the temperature remains constant or decreases outward, so there is no need for a corona. On the other hand, in that case we should also expect to find stars in which the heating varies between linearly and quadratically with density. It can be shown that such stars could neither have an extended chromosphere nor a steady hot corona (Hammer 1987a). They probably would have to undergo large scale temporal variations; perhaps they would look like hybrid stars.

Several authors (e.g., Hartmann and MacGregor 1980) noted that in magnetically open regions of low gravity red giants any type of wave tends to push rather than heat the atmosphere, because with decreasing importance of the gravity term in the momentum equation the wave pressure term becomes more important.

Castor (1981) discussed several earlier "explanations" of the dividing line and suggested that in open regions of red giants the radiative cooling time becomes smaller than the wind expansion time, so that a hot open corona would rapidly cool or not form at all. On the other hand, a hypothetical heating mechanism that operates on small temporal and large spatial scales would still be able to produce and maintain an extended hot corona. In fact, there is no reason to expect a corona to be heated only at its base. Currently discussed damping lengths for energy addition to the Solar wind and stellar winds are of the order of a stellar radius. Perhaps we should compare the radiative cooling time not with the wind expansion time, but rather with the time scale associated with fluctuations of the heat input. Then this type of argument is applicable to both open and closed regions (see below).

A number of suggestions were concerned with coronal loops. Mullan (1982), for example, suggested that on red giants beyond the dividing line closed coronal loops become dynamically unstable and are carried away by the wind. This process itself could significantly modify the wind.

Antiochos and Noci (1986) showed, as Hood and Priest (1979) had done before, that coronal loops can also exist in a "warm" state, in which thermal conduction is negligible, so that the energy balance is a local (chromospheric) one between heating and radiation (Eq. (1)). In order to avoid the need to jump to high coronal temperatures, a warm loop must fulfill two requirements: First, the maximum temperature must be smaller than the temperature at which the emission peaks (see above), and for which Antiochos and Noci take 10^5K. And second, for reasonable heating laws the loop height must be smaller than the pressure scale height at that temperature, $H_p(10^5$K). Antiochos et al. (1986) argue that as we go to low g stars, the rapid increase of the ratio H_p/R suggests that an increasing fraction of the magnetic loops can exist both in the warm chromospheric and in the hot coronal state. They further suggest that such loops prefer the warm state because, according to Antiochos et al. (1985), the hot state is linearly unstable. This effect might then explain the decay of the X-ray emission near the dividing line.

Even though this idea is certainly fascinating, a word of caution seems appropriate. The linear instability of hot loops is neither fully accepted in the literature (e.g. McClymont and Craig 1985a,b) nor particularly relevant. More important is the nonlinear behavior. Nonlinear analyses indicate either stability (e.g., Mariska et al. 1986) or limit cycle oscillations (Kuin and Martens 1982), depending on the efficiency of the mass exchange between chromosphere and corona (Craig and Schulkes 1985). (For a review on loop stability see Hammer 1987b). Moreover, the radiative cooling time of warm loops is of the order of a few seconds (depending on pressure), much smaller than the cooling time of hot loops (10^4 sec). Thus warm loops can be maintained only by a quasi-steady heating mechanism. Local fluctuations of the heat input (for example due to wave periods or time lags between microflares) are permitted only on time scales of seconds. Even if this appears to be possible on the Sun, it becomes increasingly difficult as we go to the low g stars beyond the dividing line, for which the time scales of the (sub)photospheric driving region are much larger than on the Sun. Thus, if steady warm loops really exist on such stars, the lower limit of the fluctuation time scale represents a stringent constraint for possible heating mechanisms. Otherwise, we should study highly dynamic situations.

Several other possible reasons are conceivable for why a star has no corona along a given field line, irrespective of whether this field line is open or closed. As discussed above, for any given corona model there exists a relationship between base pressure and total energy loss. Therefore, a quasi-steady hot corona can exist only if the star is able to provide the right amount of energy at the right pressure somewhere along the field line. In principle, this requirement might break down for various reasons (cf. Hammer 1987a,c). For example, the energy flux could be too small deep in the photosphere and decrease outward (due to dissipation or dilution) too rapidly with respect to the pressure. Or the pressure might be too small and/or the energy flux too large in the photosphere. Moreover, either the maximum of the emission (see above) or the apex of a loop might be reached before the coronal energy requirements can be fulfilled. In all these cases it is found that a classical Solar-like hot corona cannot exist, but only more or less extended warm regions and highly time dependent situations.

In short, we know that stars beyond the dividing line have expanding, geometrically extended chromospheres. Whether coronae become only very faint or disappear altogether is not yet fully clarified. Several reasonable suggestions have been made to account for the observations. Since many of the basic stellar properties change significantly near the dividing line, it is well possible that not only one, but several of these suggestions apply.

III. CHROMOSPHERIC HEATING

So far, the most thoroughly studied chromospheric heating mechanism has been shock dissipation of acoustic waves that are generated by the turbulent motions in the outer convection zone (Biermann 1948; Ulmschneider 1979; Jordan 1981). Recently, Kneer (1983, 1985) has suggested that such waves might also be generated in the temperature minimum region by radiative and convective instabilities associated with the formation of CO and other molecules. (For a review of such instabilities see Hammer 1987b).

The acoustic wave heating mechanism has several attractive properties, which can be
best explained in terms of a simple plane parallel wave that is generated in the
subphotosphere and propagates vertically upward. If its energy flux

$$F_M = \varrho v^2 c_s \qquad\qquad (2)$$

were strictly conserved, the velocity amplitude v would increase outward roughly
like $v \sim \varrho^{-1/2}$ because the sound speed $c_s \sim (T/\mu)^{1/2}$ varies only slowly in the
chromosphere. The larger the amplitude, the sooner can compression fronts steepen
into shocks (e.g. Hammer and Ulmschneider 1978, Eq. (5)) and dissipate the wave
energy.

A second, and even more important, energy loss mechanism of compressive waves is
due to radiative energy exchange. The details of this process depend on the properties of both the wave and the star (Ulmschneider and Kalkofen 1977). In the Sun,
radiative damping is mainly restricted to the lower photosphere, where most of the
waves have not yet formed shocks. In some other stars, shocks may form already in
the radiative damping zone, where the associated shock dissipation cannot contribute much to the energy balance. Therefore, shock heating becomes important enough
to produce a temperature inversion at the larger of the following two heights; the
shock formation height and the upper end of the radiation damping zone (Schmitz and
Ulmschneider 1981).

Above the temperature minimum, the shocks are fully developped. Their dissipation
tends to decrease the wave amplitude. This tendency counteracts the tendency of
density stratification to increase the amplitude. As a result, the amplitude becomes constant, so that according to Eq. (2) the energy flux decreases outward
roughly like the density. This property has been confirmed by detailed calculations
(Ulmschneider 1970; Schmitz et al. 1985). Therefore, acoustic shock waves exhibit
two of the main characteristics of empirical heating laws as determined from Solar
reference models: a sudden onset of the heat input and a subsequent decrease roughly proportional to the density.

Gravity waves are also produced abundantly on the Sun. They can break and dissipate
their energy in the lower chromosphere (Mihalas and Toomre 1981, 1982). Such waves
have been observed recently (Staiger 1987). Acoustic waves have also been observed;
their analysis is an important tool for probing the Solar atmosphere (Deubner 1984,
1985; Kneer and von Uexküll 1985; Deubner et al. 1987). The energy flux carried by
compressive waves is still uncertain; but in the middle and upper chromosphere
there appears to be insufficient energy to account for the heating (e.g., Deubner
1987).

Pure acoustic and gravity wave heating theories share a major flaw: they do not take the effects of magnetic fields into account. Hence, in their simplest form they are restricted to the nonmagnetic regions of the lower chromosphere. The bulk of the chromospheric emission, at least at larger heights, appears to be associated with magnetic fields.

There exists a large amount of evidence for such a connection between magnetic fields and chromospheric emission. This evidence comes both from stellar and from Solar work. Since this area has been well covered by recent reviews (e.g., in Stenflo 1983; Baliunas and Vaughan 1985; Linsky 1985a,b; Mangeney 1985; Noyes 1985; Praderie 1985; Zwaan 1986; Rodonó 1987), I will only enumerate the most important aspects. Detailed references can be found in the reviews mentioned.

First, the emission in the cores of strong chromospheric resonance lines such as CaII K varies by about an order of magnitude among stars of a given type (i.e., given T_{eff} and g) and thus given acoustic-gravity wave energy flux. The radiation from higher temperature plasma (TR and corona) varies by even larger amounts. The emissions from the various layers are well, and nonlinearly, correlated with each other (for reviews see Hammer and Linsky 1984; and Zwaan 1986). The stellar activity level is correlated with the rotation period, and to a lesser extent also with other stellar properties. Several authors have attempted to combine such dependences into dimensionless numbers that affect the dynamo, such as the Rossby number (i.e., the ratio of rotation period to convective turnover time). It is now beyond all doubt that the emission level is mostly controlled by the dynamo that generates magnetic fields. There are indications for a saturation of the emission among extremely rapid rotators, and for a minimum emission from extremely slow rotators. Only the latter minimum emission can be, and has been (Schrijver 1986), attributed to pure acoustic-gravity wave heating.

Second, only magnetic fields can account for the large scale inhomogeneity of the distribution of emitting regions over the stellar surface. Chromospheric plages, for example, are responsible for the observed rotational modulation of spectral lines such as CaIIK. By monitoring these lines over a sufficiently long time, one can infer precise stellar rotation periods, differential rotation with latitude, and the long term activity behaviour, which may be either cyclic or chaotic. If good spectral resolution and signal to noise ratio are available, it is even possible to localize individual active regions on a star and to determine their basic properties (Vogt and Penrod 1983; Gondoin 1986; Walter et al. 1987).

The Sun permits a much more detailed study of the connection between chromospheric emission and magnetic fields. Bright plages are associated with areas of strongly

enhanced magnetic flux. Even in quiet regions, the CaIIK brightness is linearly
related to the magnetic flux in the underlying photosphere (Skumanich et al. 1975).

In the quiet Sun, patches of bright CaIIK emission outline the boundaries of supergranulation cells. The resulting chromospheric network can be traced up into the
TR. Above $3 \ 10^5$K its boundaries widen rapidly, and near 10^6K the network structure
disappears. This behaviour has been interpreted in terms of the spreading with
height of magnetic flux tubes. But the large amount of fine structure observed by
instruments such as HRTS and TRC suggest more complicated geometries and prove that
the TR is highly dynamic (cf. the review by Mariska 1986). It has been suggested
that a considerable fraction of the emission formed at temperatures below $2 \ 10^5$K
might come from unresolved structures that are magnetically isolated from the
corona (Feldman 1983; Rabin and Moore 1984; Dowdy et al. 1986) such as the warm
loops discussed in Sect. II (Antiochos and Noci 1986).

It is extremely difficult to determine Solar chromospheric magnetic fields from
observations (Jones 1985, 1986). In the photosphere the field is known to be mainly
concentrated into thin magnetic flux tubes. Since the external pressure decreases
exponentially with height, these flux tubes must fan out, until they finally interfere with one another and fill all available space. The "canopy", where the mergence of field lines from neighbor flux tubes occurs, appears to lie between the
temperature minimum (above active regions) and the mid chromosphere (above quiet
regions). (Cf. Anzer and Galloway 1985; Pneuman et al. 1986; Jones 1985, 1986; and
earlier references therein.)

How can this type of magnetic structure affect chromospheric heating? To begin
with, the very presence of the magnetic field changes the physical conditions in
the nonmagnetic environment. For example, the nearly horizontal fields at the
canopy can partially reflect upward travelling acoustic waves (Abdelatif 1985).
Moreover, flux tubes exchange radiation with the ambient medium. This interaction
may generate flows and drive instabilities.

Magnetic fields can also participate actively in the transport of energy into the
upper atmosphere. The subphotospheric footpoints of flux tubes are continuously
shuffled around by granular and supergranular fluid motions. If the footpoint motions are _slow_ and chaotic, they lead to a quasi-steady rearrangement of the field
lines in a coronal loop. Parker (1982) argued that this generates a large number of
current sheets, which can dissipate by reconnection. Such reconnection processes
might become important above the canopy, where field lines merge that are anchored
in different portions of the solar surface. (Recent reviews on direct current
heating include Heyvaerts 1984; Meyer 1985; Parker 1986.)

Rapid footpoint motions lead to the generation of waves, which propagate upward along the field lines. (For complementary reviews see Stein 1985; Thomas 1985; Ulmschneider and Muchmore 1986; Roberts 1986; Hammer 1987b.) Near the Solar surface the flux tubes are thin compared with expected wavelengths (Roberts and Webb 1978; Spruit 1981); in this case three basic wave modes can be excited: Squeezing a tube produces a longitudinal (or "sausage") wave; shaking it produces a transverse Alfvén (or "kink") wave; and twisting it produces a torsional Alfvén wave. All three types of waves have been suggested to be generated at a sufficient rate to account for the heating of the Solar chromosphere (Schüssler 1984; Spruit 1984; Ulmschneider 1987).

In a strong magnetic field, the longitudinal wave is basically an acoustic wave that is channeled along the flux tube. Below the canopy the tube cross section increases with height; thus the wave energy is spread over an increasingly larger area. Consequently the wave amplitude grows less rapidly, and shock formation and dissipation occur at greater heights than in plane parallel acoustic waves of the same energy flux (Foukal and Smart 1981; Herbold et al. 1985). It has been suggested that on a given star longitudinal tube waves are generated with a much larger energy flux than nonmagnetic acoustic waves (Ulmschneider 1987). Then the larger initial value of the amplitude can offset its slower growth rate, so that these waves again produce temperature minima at about the right height.

If the field is not infinitely strong, the pressure fluctuations of a longitudinal wave cause expansions and compressions in the shape of the flux tube. The associated magnetic tension can be shown to reduce the phase speed of the wave to a value smaller than both the Alfvén speed c_a and the sound speed c_{si} inside the tube,

$$c_t^{-2} = c_{si}^{-2} + c_a^{-2} . \tag{3}$$

Under certain circumstances the pulsation of the tube walls leads to the emission of acoustic energy into the ambient medium. This energy leakage might become particularly important near the temperature minimum, where the wave amplitudes have grown large, where the tubes are no longer thin (for leaky modes in thick tubes see, e.g., Cally 1986), and where the ambient medium might be particularly cool (Ayres 1981, 1985). The cooler the external medium, the smaller is its sound speed, and the less likely are the generated waves evanescent (cf. Spruit 1982).

Transverse and torsional Alfvén waves are driven by the magnetic tension restoring force. They propagate at the Alfvén speed c_a. As far as the upward transport of energy is concerned, the Alfvén wave modes have two advantages and one disadvantage when compared with the longitudinal mode. The first advantage is that they are noncompressive as long as they are linear. Therefore, they do not suffer from radiative damping in the photosphere. The second advantage is that they are much less affected by cutoff restrictions. The cutoff frequency of the kink mode is roughly three times smaller than that of the longitudinal mode, so that it can transport energy from a much larger portion of the granular power spectrum (Spruit 1984). The torsional mode has no cutoff; it is probably excited mainly by vorticity expulsion in downdrafts (Schüssler 1984) rather than by granular motions. The disadvantage of the two magnetic modes is related to the behaviour of their phase speed c_a. Below the canopy, the Alfvén speed c_a is roughly constant, and the amplitude of upward propagating transverse and torsional waves grows at about the same rate as the amplitude of a longitudinal tube wave (i.e., with an e-folding height of about four pressure scale heights). Above the canopy, however, the magnetic field strength is roughly constant, while the density decreases exponentially. Thus, the Alfvén speed increases exponentially. As a result, the amplitudes of the transverse and torsional Alfvén waves can no longer grow, and most of their energy is reflected back. By comparison, the phase speed of the longitudinal mode increases by less than a factor of 2 within the chromosphere; major reflection occurs only at the TR.

Longitudinal waves dissipate their energy by means of shocks (cf. Ferriz-Mas and Moreno-Insertis 1987). Transverse and torsional Alfvén waves can dissipate in several different ways.

First, they can grow nonlinear and transfer part of their energy to longitudinal motions (Hollweg et al. 1982; Mariska and Hollweg 1985; Zähringer and Ulmschneider 1987), which can then dissipate in shocks. Nonlinear mode coupling should be most efficient near the canopy, where the wave amplitude is largest in terms of both sound and Alfvén speed (cf. Priest 1982).

A second possibility for dissipating Alfvén wave energy is by phase mixing (e.g., Hollweg 1981; Heyvaerts and Priest 1983). In this process, neighbor field lines oscillate out of phase. The resulting shears in the velocity and in the magnetic field strength drive Kelvin-Helmholtz and tearing mode instabilities, respectively. These instabilities can initiate a turbulent cascade of the energy down to very small scales, where it is finally dissipated by viscosity and resistivity (e.g., Hollweg 1984, 1985). Dissipation via phase mixing necessitates large velocity amplitudes; hence, it is probably most efficient near and above the canopy.

"Resonant absorption" is another important energy conversion process, in which energy from a wave propagating along a magnetic interface flows into local oscillations within the interface (see Heyvaerts 1984; Lee and Roberts 1986; and references therein).

Finally, Alfvén waves can also dissipate linearly by Joule damping, provided that the wave period is small enough (Osterbrock 1961). The resulting heating varies rather smoothly with height (Ulmschneider 1987).

Which of these mechanisms heat the Solar chromosphere? It appears likely that several mechanisms are operating simultaneously in various portions of the chromosphere. Longitudinal waves can fulfill the empirical requirements of a rapid onset of heating near the temperature minimum and a subsequent exponential decay in the lower and middle chromosphere. Transverse and torsional Alfvén waves, on the other hand, can more easily transport energy upward. They can dissipate their energy in various ways. Most of these dissipation mechanisms, however, exhibit either a smooth variation of heating with height or become most efficient only near and above the canopy, which in quiet Solar regions appears to lie substantially higher than the temperature minimum. Conversely, in active regions, where the canopy lies near the temperature minimum, all wave modes as well as reconnection can contribute to the heating of the whole chromosphere.

IV. The WILSON-BAPPU EFFECT

Modelling optically thick resonance lines such as CaIIK is a well established, although not unproblematic method to infer the structure of Solar and stellar chromospheres. A remarkable property of this line has been quantified by Wilson and Bappu (1957): The full width at half maximum, $\Delta\lambda_0$, of the emission core is well correlated with the visual magnitude of the star. In terms of fundamental stellar properties, this relationship can be written as

$$\Delta\lambda_0 \sim g^{-0.2} T_{eff}^{1.5} \qquad (4)$$

(Reimers 1973, Neckel 1974). Even more remarkable is the fact that $\Delta\lambda_0$ is nearly independent of the emission strength, and thus of the (magnetic) activity (Glebocki and Stawikowski 1978). A Solar plage region, for example, shows roughly the same width, but a much higher emission flux than a quiet region (cf. Linsky 1977). Only in sunspot chromospheres is $\Delta\lambda_0$ significantly reduced, owing to the smaller effective temperature of the umbra, but still in accordance with the Wilson-Bappu relation (Mattig and Kneer 1981). Similar width-luminosity relations have been identified for various other emission and absorption lines. Their explanation represents still a challenge and a source of controversy.

According to Ayres et al. (1975), the K_1 minima are formed in the Lorentzian damping wing of the line. This allows one to map their wavelength separation $\Delta\lambda(K_1)$

to the mass column density of the temperature minimum, and thereby to test theoretical ideas on the onset of the chromosphere. This has been done for several very different concepts: Ulmschneider et al. (1979) argued that the rapid onset of heating associated with acoustic shock waves of reasonable amplitude and period leads to the formation of temperature minima at about the right mass column densities. Ayres (1979) calculated the position of the temperature minimum for constant heating per mass and cooling by H^-. He found that the gravity dependence of $\Delta\lambda(K_1)$,

$$\Delta\lambda(K_1) \sim F_{TM}^{1/4} \, g^{-1/4},\qquad (5)$$

is consistent with the empirical relationship (4). Here F_{TM} is the total energy flux that is dissipated above the temperature minimum. Kneer (1983) argued that CO cooling may drive an instability at a position that coincides with, and perhaps even determines, the temperature minimum of the Sun. This position (mass column density) scales with g in a way that is again consistent with the empirical results.

The amazing fact that such totally different assumptions (sudden onset of heating, smooth variation of the heat input, and no mechanical flux from below) predict the same behaviour of the K_1 minima separation $\Delta\lambda(K_1)$ suggests that this behaviour is unpredictive as fas as the heating mechanism is concerned.

Under the assumption that even the $\dot{K}_2$ maxima are formed in the damping wings, Ayres (1979) calculated their separation $\Delta\lambda(K_2)$, too. He found the same gravity dependence, but the inverse heating flux dependence than for $\Delta\lambda(K_1)$. Since the Wilson-Bappu width $\Delta\lambda_0$ lies between these two quantities, the opposite F_{TM} dependences essentially cancel each other; this might be the reason for the insensitivity of the Wilson-Bappu width to the emission strength.

On the other hand, we know that the broadening of chromospheric lines is not exclusively determined by opacity, but also by nonthermal motions, in particular if such motions are taken into account in a more sophisticated way (Carlsson and Scharmer 1985) than by the usual adoption of simple micro- and macroturbulent fudge factors. From an analysis of optically thin intersystem lines of a number of stars, Ayres et al. (1982) conclude that the nonthermal velocities increase systematically with decreasing gravity. Engvold and Elgarøy (1987) find no such dependence. In any case, whether the nonthermal velocities vary with gravity or not, they do at least contribute to the broadening of optically thick lines such as CaIIK and, in particular, Hα (Cram and Mullan 1985).

This raises a very old problem: Why does this Doppler contribution to line broadening not increase with increasing stellar activity? At a first glance, one might expect that increasing heating is accompanied by increasing nonthermal velocities.

Meyer (1985) recently suggested the following solution to this (seeming) problem. It is possible that those nonthermal velocities that affect the line profiles are due to pure acoustic-gravity waves, which are constant for a given type of star. The magnetic field related heating, on the other hand, can be associated with much smaller plasma velocities if the magnetic fields are strong enough. This applies, in particular, to heating due to the annihilation of newly emerging magnetic flux.

This suggestion may well be right. However, as long as atmospheric heating is coupled to magnetic fields, an increase of the heat input can be accomplished not only by increasing the nonthermal velocities, but also by increasing the field strength or, more likely, the filling factor. In the following section, I illustrate this point by suggesting a simple model for the dependence of magnetic fine structure on global stellar properties.

V. FINE STRUCTURE vs. GLOBAL PROPERTIES

The stellar dynamo, which depends on the rotation rate and other stellar properties (Rodonó 1987; Stix 1987), produces magnetic flux. The magnetic flux erupts and protrudes through the stellar surface mainly in the form of small flux tubes. (For recent reviews on fibril fields see Parker 1986 and Schüssler 1986.) I neglect dark starspots in the present discussion.

a) linear case

Let us assume now that in a first approximation (linear case) the partitioning and energization of the flux tubes are controlled not by the dynamo (and thus by the rotation rate), but only by the physics of the photosphere and upper convection zone. In these layers the field is shredded (or merged) into tubes of a given average flux by means of instabilities and fluid motions. Then the granular and super-granular flows shuffle around, squeeze, shake, and twist the tubes; and the inter-action between flux tubes and ambient medium induces downdrafts, overstable oscil-lations, and perhaps other instabilities, both inside and outside the tube. The net effect of all these processes is that some energy flows along each tube in the form of waves and currents. Therefore, in this linear approximation both the average magnetic flux and the average total energy flux per flux tube are constant for a given type of star (i.e., given T_{eff} and g). Only the number density of the flux tubes depends on the position on the star and, via the dynamo, on the rotation rate.

Such a simple model can account for a surprisingly large number of observations. First, it can explain why the Wilson-Bappu width is so insensitive to the level of

stellar activity, independent of whether this width is determined by opacity broadening or by Doppler broadening. This insensitivity results from the fact that
within the present approximation increasing activity is caused by an increasing
number of similar flux tubes (cf. Ayres 1985); the physics of each tube, and in
particular the associated nonthermal motions, are constant for a given type of star
(i.e., given T_{eff} and g). The previous suggestions (cf. Sect. IV) that the velocities associated with magnetic fields are small (Meyer 1985) or that the K line
width is mainly dominated by opacity broadening (Ayres 1979) may be (partially)
correct for individual flux tubes; but they are unnecessary to the extent that the
activity level reflects only the number density of similar emitting objects. The
Wilson-Bappu effect proper, i.e. the gravity dependence of the K line width, could
be due to a systematic variation with gravity of either the field strength (equipartition scaling, see below), the average magnetic flux or the average total energy flux per tube, or the effects of the flux tubes on the environment.

Second, this simple picture is fully consistent with recent empirical results of
Saar and Linsky (1986, 1987), according to which the magnetic activity of a star is
controlled by the filling factor, and not by the field strength. The field strength
is found to be constant for a given star and to scale among stars in proportion to
the photospheric equipartition field.

Third, this picture is also consistent with the observed linear correlation between
CaIIK emission and photospheric magnetic flux density on the quiet Sun (Skumanich
et al. 1975) because, on the average, each flux tube on a given star carries the
same magnetic flux and the same energy.

b) nonlinear effects

Part of the magnetic flux tubes bend back towards the surface already at photospheric and chromospheric levels, but others reach larger heights and higher temperatures. Above the canopy, they must compete with each other for the limited available space.

In active regions the flux tubes are particularly abundant. When one adds more
tubes to an isolated plage region (in order to simulate a more active plage), then
the active region may partially account for the additional flux by increasing its
area in the upper layers. This effect appears to be surprisingly important
(Schrijver et al. 1985). But in the central portion of the plage, the flux tubes
are compressed when the filling factor becomes sufficiently large, and both the
magnetic field strength and the energy flux increase. This has two consequences:
First, the mergence height of field lines from neighbor flux tubes decreases; this
affects the energy transport and dissipation properties of various possible heating

mechanisms (cf. Sect. III). And second, it is reasonable to assume that with increasing energy flux in the chromosphere the coronal heating flux increases, too (see, however, the transverse Alfvén wave model by Spruit 1984). This must be accompanied by an increase of the coronal base pressure (cf. Sect. II). Indeed, according to empirical models (e.g., Basri et al. 1979; Vernazza et al. 1981; Eriksson et al. 1983) Solar plage regions and active stars tend to have higher TR pressures than quiet Solar regions and quiet stars, respectively.

For large filling factors it is conceivable that the flux tubes compete with each other not only for the available space in the chromosphere, but also for the available energy in the (sub)photosphere. Numerous flux tubes may, for example, change the character of (super)granular motions by suppressing horizontal flows; and a decreasing distance between flux tubes may affect the induced downdrafts around each tube found by Deinzer et al. (1984a,b). Clearly, all these effects can lead to a change of the types of waves and currents that are generated and to a change of the average energy flux per tube. Empirical evidence for a saturation of the relationship between chromospheric emission and photospheric magnetic flux density has been presented at this meeting by Schrijver and Coté (1987) and by Schrijver and Saar (1987).

REFERENCES

Abdelatif, T.E.: 1985, Ph.D. Thesis, University of Rochester.
Antiochos, S.K., Noci, G.: 1986, Ap.J. 301, 440.
Antiochos, S.K., Shoub, E.C., An, C.-H., Emslie, A.G.: 1985, Ap.J. 298, 876.
Antiochos, S.K., Haisch, B.M., Stern, R.A.: 1986, Ap.J. 307, L55.
Anzer, U., Galloway, D.J.: 1985, in Chromospheric Diagnostics and Modelling, ed.
    B.W. Lites, Sunspot, p. 199.
Athay, R.G.: 1981a, in The Sun as a Star, ed. S.D. Jordan, NASA SP-450, p.85.
Athay, R.G.: 1981b, Ap.J. 250, 709.
Athay, R.G.: 1985, Solar Phys. 100, 257.
Athay, R.G.: 1986, Ap.J. 308, 975.
Avrett, E.H.: 1973, in Stellar Chromospheres, eds. S.D. Jordan, E.H. Avrett,
    NASA SP-317, p. 27.
Ayres, T.R.: 1979, Ap.J. 228, 509.
Ayres, T.R.: 1981, Ap.J. 224, 1064.
Ayres, T.R.: 1985, in Chromospheric Diagnostics and Modelling, ed. B.W. Lites,
    Sunspot, p. 259.
Ayres, T.R., Linsky, J.L., Shine, R.A.: 1975, Ap.J. 195, L121.
Ayres, T.R. et al.: 1981, Ap.J. 250, 293.
Ayres, T.R., Linsky, J.L., Basri, G.S., Landsman, W., Henry, C., Moos, W.,
    Stencel, R.E.: 1982, Ap.J. 256, 550.
Baliunas, S.L., Vaughan, A.H.: 1985, Ann. Rev. Astron. Ap. 23, 379.
Basri, G.S., Linsky, J.L., Bartoe, J.-D., Brueckner, G., van Hoosier, M.E.: 1979,
    Ap.J. 230, 924.
Biermann, L.: 1948, Z. Astrophysik 25, 161.
Böhm-Vitense, E.: 1986, Ap.J. 301, 297.
Bonnet, R.M., Bruner, E.C., Jr., Acton, L.W., Brown, W.A., Decaudin, M.: 1980,
    Ap.J. 237, L47.

Brown, A.: 1986, in Cool Stars, Stellar Systems, and the Sun, eds. M. Zeilik and
    D.M. Gibson, Springer, p. 454.
Cally, P.S.: 1986, Solar Phys. 103, 277.
Carlsson, M., Scharmer, G.B.: 1985, in Chromospheric Diagnostics and Modelling,
    ed. B.W. Lites, Sunspot, p. 137.
Castor, J.I.: 1981, in Physical Processes in Red Giants, eds. I. Iben, Jr.,
    A. Renzini, Reidel, p. 285.
Cayrel, R.: 1963, Comptes Rendus 257, 3309.
Couturier, P., Mangeney, A., Souffrin, P.: 1979, in IAU Symp. 91, eds. M. Dryer, E.
    Tandberg-Hanssen, Reidel, p. 127.
Craig, I.J.D., Schulkes, R.H.S.M.: 1985, Ap.J. 296, 710.
Cram, L.E.: 1982, Ap.J. 253, 768.
Cram, L.E., Mullan, D.J.: 1985, Ap.J. 294, 626.
Deinzer, W., Hensler, G., Schüssler, M., Weisshaar, E.: 1984a,b, Astron. Ap. 139,
    426 and 435.
Deubner, F.-L.: 1984, in Small Scale Dynamical Processes in Quiet Stellar
    Atmospheres, ed. S.L. Keil, Sunspot, p. 2.
Deubner, F.-L.: 1985, in Chromospheric Diagnostics and Modelling, ed. B.W. Lites,
    Sunspot, p. 279.
Deubner, F.-L.: 1987, results presented at this meeting.
Deubner, F.-L., Reichling, M., Langhanki, R.: 1987, IAU Symp. 123, in press.
Dowdy, J.F., Jr., Rabin, D., Moore, R.L.: 1986, Solar Phys. 105, 35.
Dufton, P.L., Kingston, A.E., Keenan, F.P.: 1984, Ap.J. 220, L35.
Dupree, A.K.: 1986, Ann. Rev. Astron. Ap. 24, 377.
Dupree, A.K., Reimers, D.: 1987, in Scientific Accomplishments of the IUE,
    eds. Y. Kondo et al., Reidel, in press.
Engvold, O., Elgarøy, Ø.: 1987, poster presented at this meeting.
Eriksson, K., Linsky, J.L., Simon, T.: 1983, Ap.J. 272, 665.
Feldman, U.: 1983, Ap.J. 275, 367.
Ferriz-Mas, A., Moreno-Insertis, F.: 1987, Astron. Ap., in press.
Foing, B., Bonnet, R.-M., Bruner, M.: 1986, Astron. Ap. 162, 292.
Foukal, P., Smart, M.: 1981, Solar Phys. 69, 15.
Glebocki, R., Stawikowski, A.: 1978, Astron. Ap. 68, 69.
Gondoin, P.: 1986, Astron. Ap. 160, 73.
Gondoin, P., Mangeney, A., Praderie, F.: 1987, Astron. Ap., in press.
Hammer, R.: 1982a,b, Ap.J. 259, 767 and 779.
Hammer, R.: 1983, Adv. Space Res. Vol. 2 No. 9, p. 261.
Hammer, R.: 1987a, in Proc. IAU Symp. 122, eds. I. Appenzeller, C. Jordan, Reidel,
    in press.
Hammer, R.: 1987b, in Proc. DES Workshop, eds. E.H. Schröter et al., Cambridge UP,
    in press.
Hammer, R.: 1987c, poster presented at this meeting.
Hammer, R., Ulmschneider, P.: 1978, Astron. Ap. 65, 273.
Hammer, R., Linsky, J.L., Endler, F.: 1982, in Four Years of IUE Research,
    eds. Y. Kondo, J.M. Mead, R.D. Chapman, NASA CP-2238, p. 268.
Hammer, R., Linsky, J.L.: 1984, in Fourth European IUE Conference, ESA SP-218,
    p. 25.
Hartmann, L., MacGregor, K.B.: 1980, Ap.J. 242, 260.
Hartmann, L., Dupree, A.K., Raymond, J.C.: 1980, Ap.J. 236, L143.
Hearn, A.G., Vardavas, I.: 1981a,b, Astron. Ap. 98, 230 and 241.
Herbold, G., Ulmschneider, P., Spruit, H.C., Rosner, R.: 1985, Astron. Ap.
    145, 157.
Heyvaerts, J.: 1984, in Hydrodynamics of the Sun, ESA SP-220, p. 123.
Heyvaerts, J., Priest, E.R.: 1983, Astron. Ap. 117, 220.
Hollweg, J.V.: 1981, Solar Phys. 70, 25.
Hollweg, J.V.: 1984, Ap.J. 277, 392.
Hollweg, J.V.: 1985, in Chromospheric Diagnostics and Modelling, ed. B.W. Lites,
    Sunspot, p. 235.
Hollweg, J.V., Jackson, S., Galloway, D.: 1982, Solar Phys. 75, 35.
Hood, A.W., Priest, E.R.: 1979, Astron. Ap. 77, 233.

Jones, H.P.: 1985, in Chromospheric Diagnostics and Modelling, ed. B.W. Lites, Sunspot, p. 175.
Jones, H.P.: 1986, in Small-Scale Magnetic Flux Concentrations in the Solar Photosphere, eds. W. Deinzer, M. Knölker, H.H. Voigt, Vandenhoeck & Ruprecht, Göttingen, p. 127.
Jordan, C.: 1980, Astron. Ap. 86, 355.
Jordan, S.D.: 1981, in The Sun as a Star, ed. S.D. Jordan, NASA SP-450, p. 301.
Jordan, S.D., Avrett, E.H.(eds): 1973, Stellar Chromospheres, NASA SP-317.
Kneer, F.: 1983, Astron. Ap. 128, 311.
Kneer, F.: 1985, in Chromospheric Diagnostics and Modelling, ed. B.W. Lites, Sunspot, p. 252.
Kneer, F., v. Uexküll, M.: 1985, Astron. Ap. 144, 443.
Kuin, N.P.M., Martens, P.C.H.: 1982, Astron. Ap. 108, L1.
Lee, M.A., Roberts, B.: 1986, Ap. J. 301, 430.
Linsky, J.L.: 1977, in Solar Output and its Variation, ed. O.R. White, Colorado Assoc. Univ. P., p. 477.
Linsky, J.L.: 1980, Ann. Rev. Astron. Ap. 18, 439.
Linsky, J.L.: 1985a, Solar Phys. 100, 333.
Linsky, J.L.: 1985b, in Relations between Chromospheric-Coronal Heating and Mass Loss in Stars, eds. J.B. Zirker, R. Stalio, Sunspot, p. 55.
Linsky, J.L., Haisch, B.M.: 1979, Ap.J. 229, L27.
Linsky, J.L., Jordan, C.: 1987, in Scientific Accomplishments of the IUE, eds. Y. Kondo et al., Reidel, in press.
Lites, B.W.: 1985, in Theoretical Problems in High Resolution Solar Physics, ed. H.U. Schmidt, MPA 212, p. 273.
Mangeney, A.: 1985, Highlights of Astronomy 7, 399.
Mariska, J.T.: 1986, Ann. Rev. Astron. Ap. 24, 23.
Mariska, J.T., Hollweg, J.V.: 1985, Ap.J. 296, 746.
Mariska, J.T., Klimchuk, J.A., Antiochos, S.K.: 1986, Bull. Am. Astron. Soc. 18, 708.
Mattig, W., Kneer, F.: 1981, Astron. Ap. 93, 20.
McClymont, A.N., Canfield, R.C.: 1983, Ap.J. 265, 497.
McClymont, A.N., Craig, I.J.D.: 1985a,b, Ap.J. 289, 820 and 834.
McWhirter, R.W.P., Thonemann, P.C., Wilson, R.: 1975, Astron. Ap. 40, 63 (Erratum 1977, Astron. Ap. 61, 859).
Meyer, F.: 1985, in Theoretical Problems in High Resolution Solar Physics, ed. H.U. Schmidt, MPA 212, p. 251.
Mihalas, B.W., Toomre, J.: 1981, Ap.J. 249, 349.
Mihalas, B.W., Toomre, J.: 1982, Ap.J. 263, 386.
Mullan, D.J.: 1982, Astron. Ap. 108, 279.
Neckel, H. 1974, Astron. Ap. 35, 99.
Noyes, R.W.: 1985, Solar Phys. 100, 385.
Osterbrock, D.E.: 1961, Ap.J. 134, 347.
Owocki, S.P., Canfield, R.C.: 1986, Ap.J. 300, 420.
Parker, E.N.: 1982, GAFD 22, 195.
Parker, E.N.: 1986, in Cool Stars, Stellar Systems, and the Sun, eds. M. Zeilik and D.M. Gibson, Springer, p. 341.
Pneuman, G.W., Solanki, S.K., Stenflo, J.O.: 1986, Astron. Ap. 154, 231.
Praderie, F.: 1985, in Future Missions in Solar, Heliospheric, and Space Plasma Physics, ESA SP-235, p. 219.
Priest, E.R.: 1982, Solar Magnetohydrodynamics, Reidel.
Rabin, D., Moore, R.: 1984, Ap.J. 285, 359.
Reimers, D.: 1973, Astron. Ap. 24, 79.
Reimers, D.: 1977, Astron. Ap. 57, 395.
Roberts, B.: 1986, in Small-Scale Magnetic Flux Concentrations in the Solar Photosphere, eds. W. Deinzer, M. Knölker, H.H. Voigt, Vandenhoeck & Ruprecht, Göttingen, p. 169.
Roberts, B., Webb, A.R.: 1978, Solar Phys. 56, 5.
Rodonó, M.: 1987, these proceedings.
Rosner, R., Tucker, W.H., Vaiana, G.S.: 1978, Ap.J. 220, 643.
Roussel-Dupré, R.: 1980a,b, Solar Phys. 68, 243 and 265.

Saar, S.H., Linsky, J.L.: 1986, in Cool Stars, Stellar Systems, and the Sun, eds. M. Zeilik and D.M. Gibson, Springer, p. 278.

Saar, S.H., Linsky, J.L.: 1987, poster presented at this meeting.

Schmitz, F., Ulmschneider, P.: 1981, Astron. Ap. 93, 178.

Schmitz, F., Ulmschneider, P., Kalkofen, W.: 1985, Astron. Ap. 148, 217.

Schrijver, C.J.: 1986, in Cool Stars, Stellar Systems, and the Sun, eds. M. Zeilik and D.M. Gibson, Springer, p. 112.

Schrijver, C.J., Zwaan, C., Maxson, C.W., Noyes, R.W.: 1985, Astron. Ap. 149, 123.

Schrijver, C.J., Coté, J.: 1987, poster presented at this meeting.

Schrijver, C.J., Saar, S.H.: 1987, poster presented at this meeting.

Schüssler, M.: 1984, Astron. Ap. 140, 453.

Schüssler, M.: 1986, in Small-Scale Magnetic Flux Concentrations in the Solar Photosphere, eds. W. Deinzer, M. Knölker, H.H. Voigt, Vandenhoeck & Ruprecht, Göttingen, p. 103.

Shoub, E.C.: 1983, Ap.J. 266, 339.

Simon, T., Linsky, J.L., Stencel, R.E.: 1982, Ap.J. 257, p. 225.

Skumanich, A., Smythe, C., Frazier, E.N.: 1975, Ap.J. 200, 747.

Spruit, H.C.: 1981, in The Sun as a Star, ed. S.D. Jordan, NASA SP-450, p. 385.

Spruit, H.C.: 1982, Solar Phys. 75, 3.

Spruit, H.C.: 1984, in Small-Scale Dynamical Processes in Quiet Stellar Atmospheres, ed. S.L. Keil, Sunspot, p. 249.

Staiger, J.: 1987, Astron. Ap. 175, 263, and in preparation.

Stein, R.F.: 1985, in Chromospheric Diagnostics and Modelling, ed. B.W. Lites, Sunspot, p. 213.

Stencel, R.E., Mullan, D.J.: 1980, Ap.J. 238, 221.

Stenflo, J.O. (ed.): 1983, IAU Symp. 102, Solar and Stellar Magnetic Fields, Reidel.

Stix, M.: 1987, these proceedings.

Thomas, J.H.: 1985, in Theoretical Problems in High Resolution Solar Physics, ed. H.U. Schmidt, MPA 212, p. 126.

Ulmschneider, P.: 1970, Solar Phys. 12, 403.

Ulmschneider, P.: 1979, Space Sci. Rev. 24, 71.

Ulmschneider, P.: 1986, personal communication.

Ulmschneider, P.: 1987, Adv. Space Res., in press.

Ulmschneider, P., Kalkofen, W.: 1977, Astron. Ap. 57, 199.

Ulmschneider, P., Schmitz, F., Hammer, R.: 1979, Astron. Ap. 74, 229.

Ulmschneider, P., Muchmore, D.: 1986, in Small-Scale Magnetic Flux Concentrations in the Solar Photosphere, eds. W. Deinzer, M. Knölker, H.H. Voigt, Vandenhoeck & Ruprecht, Göttingen, p. 191.

Vernazza, J.E., Avrett, E.H., Loeser, R.: 1981, Ap.J. Suppl. Ser. 45, 635.

Vogt, S.S., Penrod, G.D.: 1983, Publ. Astron. Soc. Pacific 95, 565.

Walter, F.M., Neff, J.E., Gibson, D.M. Linsky, J.L., Rodonó, M., Gary, D.E., Butler, C.J.: 1987, Astron. Ap., in press.

Wilson, O.C., Bappu, M.K.V.: 1957, Ap.J. 125, 661.

Zähringer, K., Ulmschneider, P.: 1987, in Proc. DES Workshop, eds. E.H. Schröter et al., Cambridge UP, in press.

Zwaan, C.: 1986, in Cool Stars, Stellar Systems, and the Sun, eds. M. Zeilik, D.M. Gibson, Springer, p. 19.

SOLAR AND STELLAR CORONAE

R. Pallavicini
Osservatorio di Arcetri, Firenze, Italy

ABSTRACT. This paper reviews our present understanding of coronal heating, structuring and variability that has emerged from space observations of the Sun and nearby stars. It is shown that a basic analogy exists between solar coronal physics and the phenomena observed in the coronae of other late-type stars. Recent X-ray observations of stellar coronae from EXOSAT are used to illustrate the main points.

1. INTRODUCTION

In this paper, I will not attempt to present a comprehensive review of solar and stellar coronal physics. The subject has developed so enormously in the past few years that to cover it all has become a virtually impossible task. Instead, I will focus on a few general topics which I feel are particularly important and suitable to illustrate the many contact points between solar and stellar coronal studies. In particular, I will discuss the following fundamental problems:

  a) the heating of solar and stellar coronae, both in the framework of dynamo models and in the light of recent suggestions which relate quiescent emission to flaring activity in stars;

  b) the structuring of coronae, as derived directly from spatially resolved solar observations and indirectly from modelling of spatially unresolved stellar observations;

  c) the variability of solar and stellar coronal emission, particularly the most obvious manifestations of such variability, i.e. the occurrence of flares at coronal heights.

The basic observational data we need to study the above problems have nearly all been provided by space missions that in the past several years have monitored the X-ray and UV emission from the Sun and nearby stars. In particular, observations of the Sun from SKYLAB, OSO-8, SMM and other satellites have demonstrated the essential role played by magnetic fields in shaping coronal structures and in controlling the flow of mass and energy. At the same time, stellar observations with IUE, EINSTEIN and EXOSAT have shown that transition regions and coronae are ubiquitous among stars of almost all spectral types. It is quite obvious at present that solar and stellar observations are mutually complementary and need to be analyzed in toto in order to solve the basic physical problems of coronal physics. This is a point that I will keep emphasizing in the course of this review.

2. CORONAL HEATING

As it is well known, in the outer solar atmosphere the temperature first drops to a minimum in the upper photosphere and then starts rising again reaching $\approx 10^4$ K in the chromosphere and more than $\approx 10^6$ K in the corona. This apparent paradox has been a major puzzle in solar physics since the discovery of the high coronal temperature in the 1940's. The theory almost universally accepted up until now explained the temperature inversion as due to shock dissipation of acoustic waves generated in the subphotospheric convective zone. In its simplest formulation, the theory neglected magnetic effects and regarded magnetic activity as a localized phenomenon with only minor effects on the global properties of the Sun. In such a scenario, the "average" corona outside activity centers could only be expected to be quite homogeneous and temporally constant.

SKYLAB showed a completely different picture: the corona appeared to be formed by an ensemble of arch-shaped structures (loops), connecting regions of opposite magnetic fields (Vaiana and Rosner 1978). These loops differ widely according to temperature, density and length; the more extended loops usually have a lower temperature and density than the shorter ones. These structures are ubiquitous in the solar corona and there is no evidence of a quiet homogeneous substratum in the space between them. Loop structures occur in bright points, active regions, flares and even in the "quiet" low magnetic field regions outside centers of activity. The only exception are coronal holes, i.e. regions of lower temperature and density where the magnetic field opens into the interplanetary medium and from which the high velocity wind originates. Since the solar corona is a highly conductive, low β plasma, the loops actually trace magnetic field lines: the heating appears to occur in magnetically confined structures and is strongly enhanced in regions of high photospheric magnetic field. Although SKYLAB could not prove that the solar corona is heated by magnetic processes, it pointed out the importance of considering magnetic effects and motivated a search for alternative heating mechanisms based on dissipation of MHD waves and/or DC currents (Kuperus, Ionson and Spicer 1981, Parker 1983).

Perhaps the most important indication that magnetic effects are important for heating the solar corona was provided by OSO-8 and SMM observations (Athay and White 1978, Bruner 1978, 1981). By measuring Doppler broadenings of UV lines formed in the transition region it was possible to demonstrate that the energy flux carried by acoustic waves was too short by several orders of magnitude to provide for the energy requirements of the transition region and the corona, although it could be high enough to heat the low chromosphere. Additional evidence in favor of magnetic processes came from the virtually continuous time variability of transition region and coronal emission. This variability appears to be stochastic in nature and indicates the existence of continuous fluctuations in the heating rate, that are probably associated with the emergence of new magnetic flux or with the stressing of magnetic field lines by turbulent surface motions.

What can stellar observations tell us about the heating problem? The extensive observations of X-ray emission from stellar coronae carried out by the EINSTEIN Observatory showed the existence of severe discrepancies in the acoustic theory of coronal heating and called for different explanations. In order to have an idea of the situation before EINSTEIN, let us imagine a diagram of predicted and observed X-ray coronal emission vs. spectral type as could be drawn just before the launch of EINSTEIN (a similar diagram appears as Fig. 5 in Mewe 1979). The stellar X-ray luminosities were calculated on the basis of the dissipation of acoustic waves generated in subphotospheric convective zones. Only stars in a narrow range of spectral types, therefore, were predicted to have coronae; not surprisingly, these were mostly stars of spectral types F and G, i.e. stars not very different from the Sun. In the stan-

dard theory, stars of spectral types earlier than F were not expected to have coronae
simply because they do not possess outer convective zones. Stars of very late spectral types (K and M), although possessing deeper convection zones than the Sun, were
not expected to be vigorous X-ray emitters since the generation of the acoustic flux
scales as the eigth power of the convective velocity, which decreases towards later
spectral types. For the earliest spectral types (O and B stars), for which UV emission was already known, a somewhat different theory had to be proposed, based on the
amplification of acoustic waves by the radiation pressure of these high luminosity
stars. In the same diagram, one could also plot the handful of stellar coronal
sources known at that time (see again Fig. 5 in Mewe 1979). Notice that some of
these sources have not been confirmed by later observations.

The EINSTEIN Observatory showed that in reality the story was completely different (Vaiana et al. 1981, Rosner, Golub and Vaiana 1985). Stars of virtually all
spectral types were found to be X-ray emitters with luminosities ranging from
$\approx 10^{26}$ erg s^{-1} to $\approx 10^{34}$ erg s^{-1}. The only exceptions were A-type dwarfs - for
which there is no credible evidence of X-ray emission at the present sensitivity
level - and late-type giants and supergiants. In particular, the EINSTEIN Observatory showed that early-type stars (O and B) are very strong X-ray sources, with
luminosities in the range of $\approx 10^{29}$ erg s^{-1} to $\approx 10^{34}$ erg s^{-1}, much larger than previous
predictions by even the most optimistic theories. For late-type stars (F to M) there
is a broad range (up to three orders of magnitude!) of observed X-ray luminosities
at each spectral type, but there is little, if any, dependence of the median X-ray
level on spectral type. These observations are in obvious contrast with the standard theory of coronal formation by dissipation of acoustic noise. We note in passing that among the detected sources there was a group of very active late-type stars
with typical X-ray luminosities in the range $\approx 10^{30}$ to 10^{31} erg s^{-1}. These were
binaries of the RS Canum Venaticorum type, which were established as a class of
X-ray sources by early observations with HEAO-1 and which are now recognized as
being the brightest X-ray sources among late-type stars (I will come back to these
stars later on).

The EINSTEIN Observatory also showed that there is a dichotomy between early-
type stars (O and B) and late-type stars (F to M) regarding the dependence of
X-ray coronal emission on basic stellar parameters (Pallavicini et al. 1981). While the
X-ray luminosity of early type stars depends on bolometric luminosity ($L_X \sim L_{bol}^{-7}$)
and is virtually independent on rotation, the reverse occurs for late-type stars
which show no dependence on the radiation field but a strong dependence on rotation
($L_X \sim v_{rot}^2$). This remarkable dichotomy suggests that, whatever the heating mechanism
of coronae, it must be fundamentally different for early- and late-type stars. For
the latter ones, we can probably use the solar analogy as a guide-line; for early-
type stars, instead, a similar approach is likely to be completely unjustified and
may lead to erroneous conclusions. As I will show in a moment, stellar observations
of late-type stars are consistent with magnetic heating of the coronal plasma, as
independently suggested by solar observations. For early-type stars, no satisfactory
theory exists as yet, although some plausible scenario, which needs to be tested by
future X-ray missions, has already been proposed (e.g. Lucy and White 1980).

For stars of late spectral types, the observed dependence of coronal emission
on rotation can be taken as a strong argument in favor of the notion that coronal
heating is due to dynamo generated magnetic fields emerging at the star surface. This
is further supported by the rapid onset of X-ray emission among late-type stars at
spectral type $\approx$ F0, i.e. at about the same spectral type at which stars start to have

appreciable outer convective zones. This rapid onset contrasts with the sharp drop
off of X-ray emission at spectral type A (Schmitt et al. 1985a). Amplification of
magnetic fields by dynamo action requires interaction of rotation and convection,
which in turn produces differential rotation. Differential rotation acting over a
seed poloidal field produces a toroidal field which eventually emerges at the star
surface by magnetic buoyancy (see reviews by Cowling 1981 and Gilman and DeLuca 1986).
The efficiency of dynamo action is expected to depend on both rotation and convection,
in qualitative agreement with present X-ray observations. Unfortunately, it is difficult to go beyond this qualitative stage. Dynamo theories are not sufficiently well
developed, even in the case of the Sun, to allow reliable predictions to be made for
other stars. Moreover, it is important to keep in mind that the magnetic field
generated by the dynamo process is related only indirectly to the observed X-ray
coronal emission. There are many steps in between, all poorly understood, including
the concentration of the field in elementary KGauss flux tubes, the emergence at the
star surface, the stressing of magnetic field lines by turbulent fluid motions, the
mechanism of conversion of magnetic energy into thermal energy and plasma heating.
Only when all these processes are clearly understood for the Sun, it will be possible
to relate quantitatively the observed X-ray emission of stellar coronae to basic parameters of the dynamo process, such as rotation and convection. For the moment, we
must be content with the good qualitative agreement found between observations and
general expectations of the dynamo theory.

Now I would like to discuss a new and interesting suggestion for heating the
solar corona and the coronae of other late-type stars. Using balloon-borne high
sensitivity instrumentation, Lin et al. (1984) have found that small amplitude hard
X-ray bursts, that they have called "microflares", occur frequently on the Sun.
These events, which typically occur with a frequency of one every 5 minutes or so,
have an intensity 10 to 100 times smaller than all previously detected hard X-ray
bursts and tend to occur during the rise phase of the associated soft X-ray events,
as usually observed in solar flares. Taking into account their rate of occurrence,
one can estimate that the average rate of energy deposition by greater than 20 KeV
electrons is $\simeq 10^{24}$ erg s^{-1} . This is quite small with respect to the energy required to heat the corona ($\simeq 10^{27}$ erg s^{-1}). However, the available energy may be
substantially higher if the electron spectrum extends to lower energies, for instance
as low as 10 or 5 KeV. Moreover, the rate of occurrence of microflares continues to
increase as the detection threshold decreases. This opens up the possibility that
the energy deposited by non-thermal electrons may indeed be sufficient to heat the
solar corona. There are good theoretical reasons to believe that coronal heating
may actually occur in discrete events through the dissipation of current sheets
formed as a consequence of the shuffling of magnetic footpoints by fluid motions
(Parker 1983).

The concept of coronal heating as a result of continuous flaring activity has
been suggested also in the stellar case, with some important differences with respect
to the solar one. This suggestion is based on two main arguments. First, it has
been found by a number of authors (Doyle and Butler 1985, Skumanich 1985, Whitehouse
1985) that there is a statistical correlation between the time averaged rate of energy release in optical U-band flares and the quiescent X-ray luminosity of dMe stars.
What is even more important, there appears to be an approximate equality between the
total energy released by flares (averaged over time) and the X-ray quiescent luminosity of dMe stars (Doyle and Butler 1985). A relationship between quiescent and flaring activity is not unexpected if they both originate from the same basic physical
mechanisms, such as stressing and dissipation of dynamo generated magnetic fields.

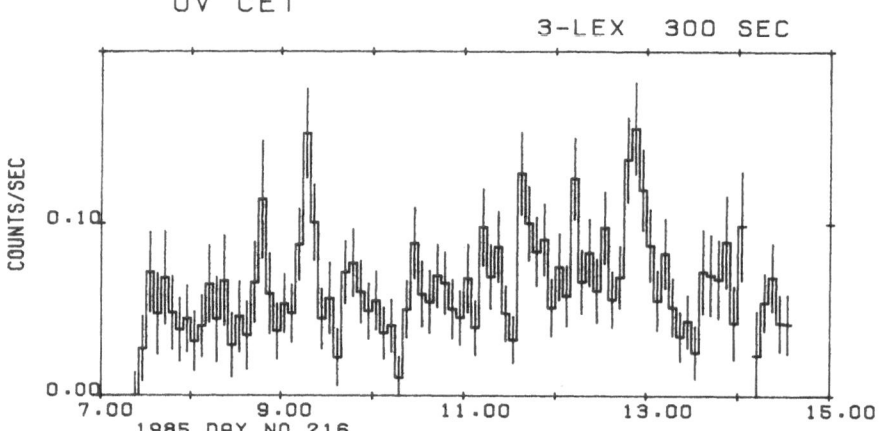

Fig. 1 - EXOSAT LE observation of UV Cet on August 4, 1985 (from
Pallavicini and Stella 1987).

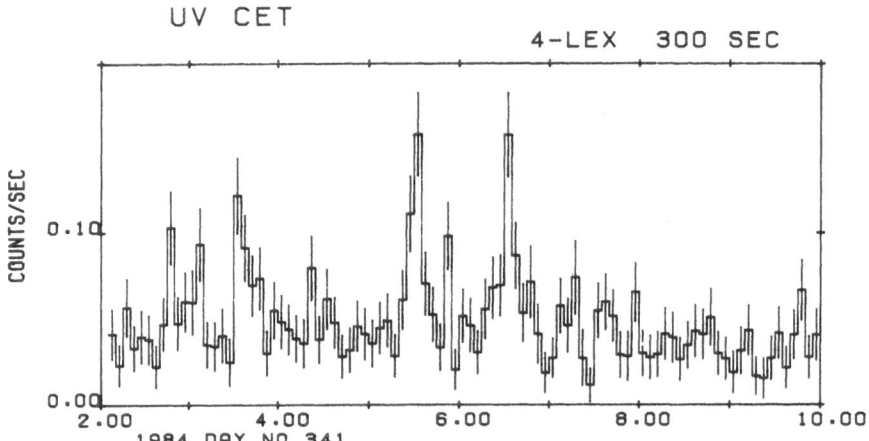

Fig. 2 - EXOSAT LE observation of UV Cet on December 6, 1984 (reanalyzed data from the EXOSAT Archive). Only the period
02:00 to 05:00 UT was covered by simultaneous optical
spectroscopy as reported by Butler et al. (1986).

However, the conclusion that flares energize stellar coronae would require the very special condition that flares deposit in the atmosphere (for instance by mass motions) an equal amount of energy to that emitted by them in electromagnetic radiation. How this non-radiative energy can be stored and subsequently reabsorbed by the atmosphere remains to be explained. The second argument is more direct: Butler et al. (1986; see also Butler and Rodonò 1986) have claimed to have found evidence of microflaring activity in EXOSAT observations of flare stars obtained simultaneously with optical spectroscopy at ESO. They base this conclusion on an observation of the star UV Cet obtained in December 1984. The EXOSAT data have been binned by them at very short time intervals (30 sec and 60 sec), in spite of the low intensity level of the source ($\approx$ 0.05 counts per second). They notice the "spiky" appearance of the data, which at a first glance looks just like noise. However, simultaneous monitoring of the same star in the Hγ line shows that some of the X-ray peaks are clearly associated with brightenings in Hγ, which suggests that they might be statistically significant (see Fig. 1 in Butler et al. 1986). They conclude that the quiescent corona of dMe stars may actually result from a continuous succession of microflares lasting from tens of seconds to several minutes and with characteristic energies of $\approx$ 2 X 10^{30} erg.

Given the importance of this result for solar and stellar physics, we have undertaken an extensive survey of flare star observations obtained with EXOSAT. There are about 20 such observations in the EXOSAT archives, each of them lasting normally for periods of $\approx$ 8 hours. The data have been analyzed for time variability, using a technique developed by L. Stella, which compares the observed variance with the variance expected from a constant source, taking into account the bias introduced by the fact that a counter was used in acquiring the data. In addition, an auto-correlation analysis is applied which allows the determination of the relevant time scales of shot-noise fluctuations. Fig. 1 shows the best case for continuous variability that we have been able to find (Pallavicini and Stella 1987). It is an observation of UV Cet obtained several months later than the previous observation by Butler. The star is clearly variable at a high significance level ($\approx$ 7σ). The observed variability is mainly due to three flares detected at a greater than 4σ significance level (at 09:20, 11:40 and 12:50 UT). Other smaller events are also probably present, at a significance level not exceeding 3σ. The observed flares last much longer (tens of minutes) and have higher energies (from $\approx$ 5 X 10^{30} to $\approx$ 1 X 10^{31} erg) than the microflares reported by Butler. Furthermore, the source is a visual binary unresolved by EXOSAT, and there is no reason why the observed events should all originate from the same star. In fact, simultaneous radio observations at the VLA by Kundu et al. (1987) clearly demonstrate that the three major events occurred alternatively on both components of the system. We have also reanalyzed in the same way the original observation by Butler (Fig. 2). We find that the source behaved quite similarly during the two observations, with regards to both quiescent and flaring emission. We only found evidence for the occurrence of relatively major events, lasting five to ten minutes each, with no significant variability detected on time scales as short as $\approx$ 1 min or smaller. During the period of simultaneous optical spectroscopy (02:00 to 05:00 UT), the larger events were associated with Hγ peaks, but there were also many other Hγ peaks not associated with simultaneous X-ray brightenings, which is hardly surprising considering what we know from the Sun. In addition, the last three hours of the observation by Butler are completely free of significant variability, which is difficult to reconcile with the interpretation of quiescent emission as a superposition of low amplitude flares.

All other observations analyzed by us lead to the same conclusion. We find substantial variability in dMe flare stars on all time scales from several minutes to hours. However, we do not find any convincing evidence that quiescent emission on these stars may result from continuous low-amplitude and short-lived flaring activity. Two other EXOSAT observations of UV Cet may be particularly relevant (one is shown in Fig. 5 later on). Excluding a major flare, the source was observed at about the same flux level as on previous occasions: however, the observed variability was quite different, and dominated by long time scales (of the order of $\simeq 1/2$ hour to 1 hour) in addition to variability on time scales of $\simeq 15$ min. Again there was no evidence of significant variability on time scales of $\simeq 1$ min or shorter. In conclusion, we cannot exclude at this stage that microflaring activity might heat the coronal plasma. However, in our opinion, this is still somewhat speculative in the case of the Sun, and, what is more, it remains to be proven in the case of stars.

3. CORONAL STRUCTURING

On the Sun, virtually all of the observed X-ray emission originates from loop structures, which are the building blocks of the solar corona. A loop structure is simply a magnetic flux tube with rigid walls, inside which high temperature, high density plasma is confined. If we neglect a number of distracting complications (e.g. departure from symmetry, variations of loop cross-section with height etc), we can characterize a loop structure with a few global parameters, such as semilength L, maximum temperature T_M at the loop top, and base pressure p_0 (equal to the pressure along the entire structure for loops much smaller than the pressure scale height). The simplest way of modelling such a structure is to assume static conditions (i.e. $v = 0$ everywhere) and to impose at each point in the loop an energy balance equation between the heating rate E_H per unit volume, the radiative losses E_R (assumed to be optically thin), and the divergence of the conductive flux F_C, i.e.

$$E_H + E_R = \text{div } F_C \tag{1}$$

where $F_C = K \, T^{5/2} \, (dT/dl)$. The conductive flux will vanish at the loop top (for symmetry reasons) and at the loop footpoints, provided these are taken at sufficiently low levels in the atmosphere that the temperature gradient at the footpoints is virtually zero (in practice, we take the loop base in the chromosphere at the level where $T \simeq 2 \times 10^4$ K). With these boundary conditions, Eq. (1) implies that the total energy deposited in the loop by the heating mechanism must be balanced by the total radiative losses, conduction being only a means of transferring energy from one part of the loop to another.

With the aid of Eq. (1) it is possible to obtain simple scaling laws between global parameters of the loop. For instance, in the case of constant pressure and constant heating deposition along the loop, the scaling law reads (Rosner, Tucker and Vaiana 1978):

$$T_M \sim (P_0 \, L)^{1/3} \tag{2}$$

Alternatively, we can write:

$$T_M \sim E_H^{2/7} \, L^{-4/7} \tag{3}$$

Eq. (2) has been shown to be in reasonable agreement with observations of spatially resolved features on the Sun. On the Sun we also observe that, except for flares, the coronal temperature does not vary very much for different types of structures, and that the high temperature, high density loops, typical of active regions, are shorter than the low temperature, low density loops characteristic of quiet areas.

When we observe the integrated X-ray emission from a stellar corona, all the different structural features will be mixed together, which makes the analysis much more complicated. If we can assume that the corona is constituted by structures of only one type, all with the same pressure and length (an assumption which at best is only a crude approximation of reality), we may expect that the observed emission will depend on many parameters, including the number of loops, their length and base pressure, their cross-sectional area, the distribution of temperature along the structure, and so on. However, as shown by the existence of the scaling laws above, not all these parameters will be independent. In practice, we can express the integrated coronal emission from a star as:

$$F_X \sim \Psi \ (T_M, \ P_0, \ A_f) \tag{4}$$

where A_f is the fraction of the stellar disk covered by X-ray emitting loop structures, and Ψ is a function which can be computed (generally numerically) on the basis of the energy balance equation (1). The relevant question is whether X-ray observations of stars are capable of providing us with a determination of the three parameters T_M, P_0 and A_f, or at least can allow us to put constraints on these quantities. Before we answer this question, we need to discuss briefly what the observations are actually telling us.

An important physical quantity that we would like to know is the temperature stratification in stellar coronae. In order to obtain this information we need high spectral resolution observations, which so far have been obtained only for a limited number of bright nearby sources, mostly RS CVn and Algol-type binaries. Fig. 3 shows an observation of Capella obtained with the Transmission Grating on EXOSAT (Mewe et al. 1986). Although we cannot resolve individual lines formed at different temperatures, the complexes of lines resolved by the instrument already call for a two-temperature model, with one component at a temperature of a few million degrees, and the other one at a temperature of about one order of magnitude higher. Two temperature models were also required by previous observations of RS CVn and Algol-type binaries obtained with the SSS instrument on board the EINSTEIN Observatory (Swank et al. 1981). Also low-resolution EINSTEIN IPC data, as well as EXOSAT broad-band observations of both RS CVn's and normal active stars, usually give two temperature solutions (Majer et al. 1986, Pallavicini et al. 1987). Taken at face value, this result, which suggests the existence of spatially distinct regions at quite different temperatures, is at variance from what we would expect from the solar analogy. On the Sun, even if we consider it as a mixture of quiet and active regions, the coronal temperatures do not differ by more than a factor of ≈ 2; furthermore, on the Sun, we have a continuous distribution of temperatures inside coronal structures, rather than isothermal regions in two quite distinct temperature regimes.

J. Schmitt was probably the first one to notice a curious behavior when comparing results obtained with different instruments (Schmitt 1984, see also Majer et al. 1986). He noticed that, when observing different stars with the same instrument, one usually gets very similar temperatures; on the contrary, when the same stars are observed with different instruments, the derived temperatures usually do not particularly agree. This is well illustrated by a comparison of spectral fittings of EINSTEIN SSS and IPC spectra of RS CVn stars (see Figs. 3 and 4 in Majer et al. 1986). This behavior is what could be expected from a continuous distribution of temperatures in stellar coronae. As any observing instrument has a finite pass-band, it will select those plasma regions which are at temperatures at which the instrument is most sensitive. In other words, the temperatures we can measure are not "true" coronal temperatures, but rather "effective" temperatures, which depend on both the differential emission measure distribution in the source and the spectral response of the instrument used.

In order to test this hypothesis, Schmitt et al. (1987) have made MonteCarlo simulations of EINSTEIN and EXOSAT coronal observations, for sources with a continuous differential emission measure distribution of the form $\sim T^\alpha$ (which closely mimics that of a coronal loop in energy balance). For the IPC they obtain the typical two temperature solutions usually found in real IPC spectra of coronal sources; for EXOSAT broad-band observations, they find a distribution of filter ratios Al-Pa/3-Lex which clusters around values ≈ 0.5, similar to the observed values. Notice that EXOSAT filter ratios do not provide unique temperature solutions, even for truly isothermal sources. A detailed analysis of EXOSAT broad-band results shows indeed that filter ratios obtained with different filters can only be reconciled by assuming a continuous temperature distribution in the source (Pallavicini et al. 1987).

We can now come back to Eq.(4) and to the determination of source parameters from observations. It is clear from the above discussion that the maximum temperature T_M in the loop is not a quantity that can be directly measured. The observations can only provide an "effective" temperature which will usually be lower than T_M. Nor can we observe directly P_0, since it would require the measurement of density sensitive lines, a task that needs X-ray instruments with much higher spectral resolution than those flown so far. However, if we have an X-ray spectrum or measurements at different X-ray and UV wavelengths, we can hope to get some information on source parameters by model fitting techniques. This has been done by a number of authors, with different degrees of success. Broadly speaking, we can distinguish two different types of approach. Certain people (Schmitt et al. 1985b, Stern, Antiochos and Harnden, 1986) try to fit X-ray observations only, i.e. the coronal portion of the loop; others (Giampapa et al.1985, Landini et al. 1985, Landini, Monsignori-Fossi and Pallavicini 1985) try to find a consistent solution for X-ray and UV observations (as provided by IUE) on the grounds that a loop model should be able to simultaneously reproduce the coronal portion of the loop, as well as the transition region and chromospheric sections at the base of the same structure. Generally speaking, the first approach usually gives reasonably good results (i.e. loop models are able to reproduce X-ray observations); much more difficult is to find self-consistent solutions for the entire set of X-ray and UV observations. Whether this is due to temporal variability between X-ray and UV observations of the same stars, or to the existence on stars of a family of cool loops with maximum temperature not exceeding $\approx 10^5$ K - in addition to the more familiar hot loops (cf. Antiochos and Noci 1986) - this is not understood. We note in passing that even the coexistence on the same star of coronal loops in quite different physical conditions (as may occur for RS CVn binaries, see below) may result in poor agreement between the observations and the predictions of single loop models.

In spite of the many uncertainties that still plague loop modelling of stellar coronae, a few facts already seem well established. First, the maximum temperatures T_M derived for many active late-type stars (both single and of the RS CVn type) are

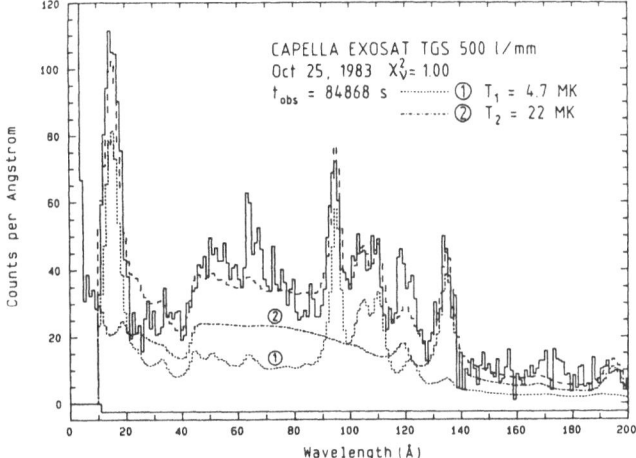

Fig. 3 - EXOSAT Transmission Grating observation of
Capella (from Mewe et al. 1986).

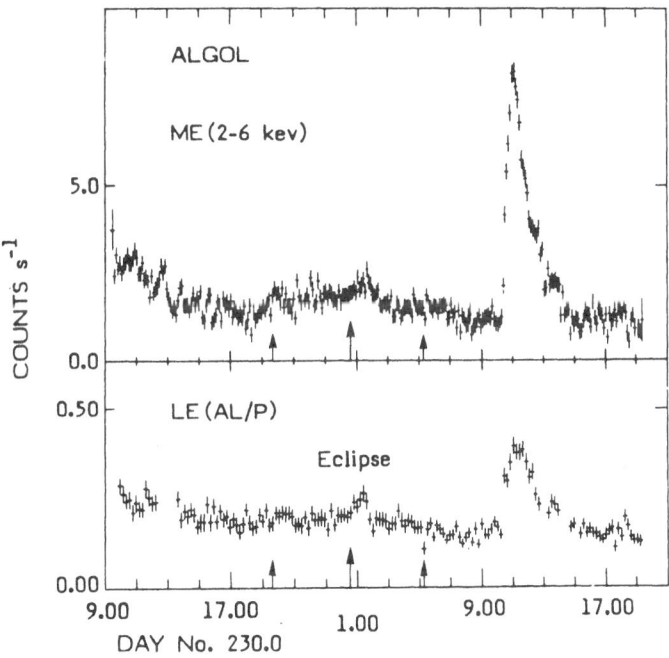

Fig. 4 - EXOSAT LE and ME observations of Algol centered
around the phase of the secondary eclipse. A large
flare was also detected (from White et al. 1986).

far in excess than those typical of solar active regions and approach those of the flaring Sun ($\approx$ a few times 10^7 K). If we recall the scaling law (3), this means that, for loops of comparable length, the heating rate must be larger. Actually, since E_H scales as $T_M^{7/2}$, the heating rate must be many orders of magnitude higher in active late-type stars than in the Sun. The other important result is that it is not possible to separate the area filling factor A_f from the base pressure P_0, unless additional constraints are imposed quite arbitrarily. This is simply because the observed X-ray emission depends linearly on the amount of emitting coronal plasma and there is no obvious way of distinguishing, in spatially integrated observations, between large low pressure regions and small high pressure regions.

Fortunately, there is another way by which we can get information on the spatial structure of stellar coronae. This is through eclipse observations in eclipsing binary systems as well as through rotational modulation of the observed emission in rapidly rotating single stars. This technique was first applied by Walter, Gibson and Basri (1983) using the EINSTEIN Observatory. They observed the RS CVn binary AR Lac, which is formed by a G2 IV primary and a K0 IV secondary, separated by 9.1 R_0. In spite of the many gaps present in the data owing to the low orbit of the satellite, they were able to observe a deep primary eclipse (when the G star is occulted by the K star) and a shallow secondary eclipse (when the K star is behind the G star). From this they concluded that compact coronal structures exist on both stars and, in addition, that an extended inhomogeneous corona surrounds the K0 IV component. However, it was only with the long, uninterrupted observations provided by the EXOSAT satellite that it has become possible to fully exploit this technique (White et al. 1986, White 1987). EXOSAT has obtained observations of several systems, including a 35 hours continuous observation of Algol, centered on the secondary eclipse, and complete coverage of a full orbital period for AR Lac (P=1.98 days) and TY Pyx (P=3.20 days). The observation of Algol failed to reveal any eclipse when the K0 IV X-ray bright component was behind the X-ray dark B8 V primary (Fig. 4). From this it was inferred that an extended high temperature corona with a scale height of at least $\approx$ 1 R_* surrounds the K star. The observation from AR Lac is even more interesting (White 1987). The primary eclipse was observed in the Low Energy detector (as it was with EINSTEIN), but there was no obvious eclipse in the Medium Energy data: this indicates that the extended corona is at higher temperatures than the more compact structures close to the surface of the star.

The picture which emerges from these eclipse observations indicates that structures of quite different sizes, temperatures and pressures coexist in RS CVn binaries. The more compact structures close to the star surface are at lower temperature and probably higher density than the more extended structures whose sizes are comparable and even larger than the stellar radius. What are these extended components? We do not know. They might be associated with loops connecting the two stars, as suggested by extrapolations of photospheric magnetic fields (Uchida and Sakurai 1983) and further supported by VLBI observations which indicate the presence of extended radio emitting components comparable in size with the binary separation (Mutel et al. 1985, Felli et al. 1987). Alternatively, the extended components may be the upper end of a range of solar-like loops whose larger dimensions are permitted by the lower gravity of RS CVn stars. It is perhaps interesting to observe that the different average temperatures found for compact and extended structures in RS CVn binaries repropose the question of the two-temperature solutions found from fitting EINSTEIN and EXOSAT spectra. Schmitt (1987), working in the context of coronal loop modelling, has already found it difficult to explain the extended coronal structures suggested by eclipse observations. The question whether RS CVn binaries behave similarly to the

Sun and other late-type stars, or rather if they possess unique properties, should
be considered open at this stage.

4. CORONAL VARIABILITY

Surprisingly, very little was known until quite recently on time variability of
coronal X-ray sources. The observations from the EINSTEIN Observatory were usually
very short ($\approx$ a few thousands seconds) and longer observations - when available -
were interrupted by the periodic eclipses of the satellite and by the passages
through regions of high background. Furthermore, systematic effects and the poor
knowledge of source temperature make it difficult to compare observations obtained
with different satellites. For instance, fluxes measured by EINSTEIN and EXOSAT may
have systematic differences of up to a factor of $\approx$ 2, owing to the different spectral bands and to the temperature dependence of the detector response. The most
comprehensive study of stellar X-ray variability before the advent of EXOSAT is probably that of Ambruster, Sciortino and Golub (1987). They have analyzed EINSTEIN
observations from a sample of active late type stars (mostly flare stars), using a
new substantially improved version of the classical χ^2 test. They have found that
variability is ubiquitous in their sample of K and M stars, with a typical amplitude of $\approx$ 30% and time scales ranging from a few hundred seconds to > 1000 sec.
However, the presence of many data gaps, characteristic of EINSTEIN observations,
makes the physical nature of the observed variability rather unclear. In addition,
long time scales (including the entire evolution of long-lived transient events)
could not be adequately studied with EINSTEIN.

The EXOSAT satellite has dramatically increased our knowledge of time variability of stellar coronae, at least at short and medium time-scales (minutes to days).
I have already mentioned above some of the EXOSAT results on time variability. In
this section, I will focus only on one aspect of variability, that associated with
flare-like events. I will show that with the new EXOSAT observations we are starting
to appreciate the rich variety of transient phenomena observed on stars, and we are
in a better position to make comparisons with similar phenomena observed on the Sun.
The quality of the data is such as to allow for the first time realistic modelling
and the derivation of miningful physical parameters.

Flares have been observed by EXOSAT on a variety of stars (Brinkman et al. 1984,
Droemer and Gibson 1985, de Jager et al. 1986, Landini et al. 1986, Pallavicini et
al. 1986, White et al. 1986, Nelson et al. 1987, Haisch et al. 1987, Pallavicini and
Stella 1987). They have been observed on classical dMe flare stars (UV Cet, AT Mic,
YZ CMi, EQ Peg, YY Gem etc.; cf. Figs. 5 and 6), on RS CVn and Algol-type sytems
(σ^2 CrB, Algol; cf. Fig. 4) and even on a single solar-type G star (π^1 UMa, cf. Lan-
dini et al. 1986). The latter observation may not be particularly surprising, since
X-ray variability and flares are commonly observed in integrated observations of the
Sun. What is surprising, however, is that the flare on π^1 UMa was seen against a
background quiescent emission that was two orders of magnitude higher than the solar
quiescent luminosity! This implies that the π^1 UMa flare released, in the X-ray
band alone, at least a factor ~ 10 more energy than the total energy released by
the largest solar flares. This suggests that activity on young rapidly-rotating
solar-type stars (such as π^1 UMa) does not result simply from larger areas of the
stellar surface covered by magnetic regions: it may also be intrinsically more
powerful. An even more interesting observation is that shown in Fig. 7. It is a
3-Lex observation of the bright star Castor (α Gem), which is formed by two unresol-

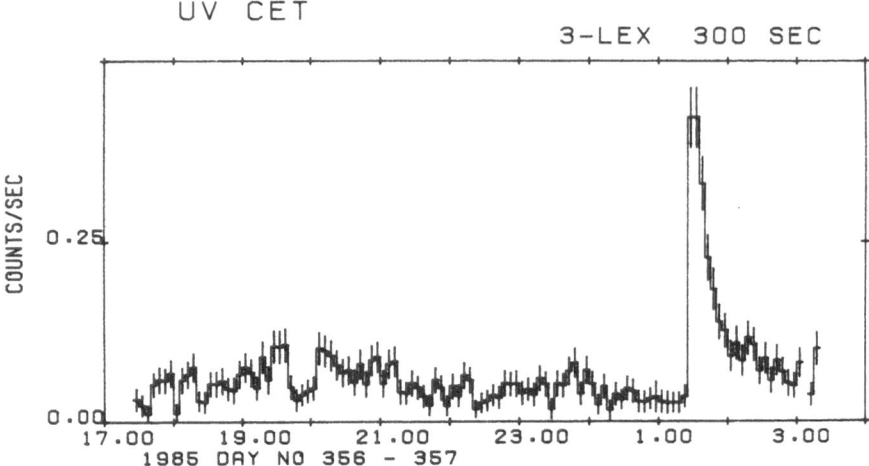

Fig. 5 - EXOSAT LE observation of a large impulsive flare on UV Cet on
December 23, 1985 (unpublished data from the EXOSAT Archive).

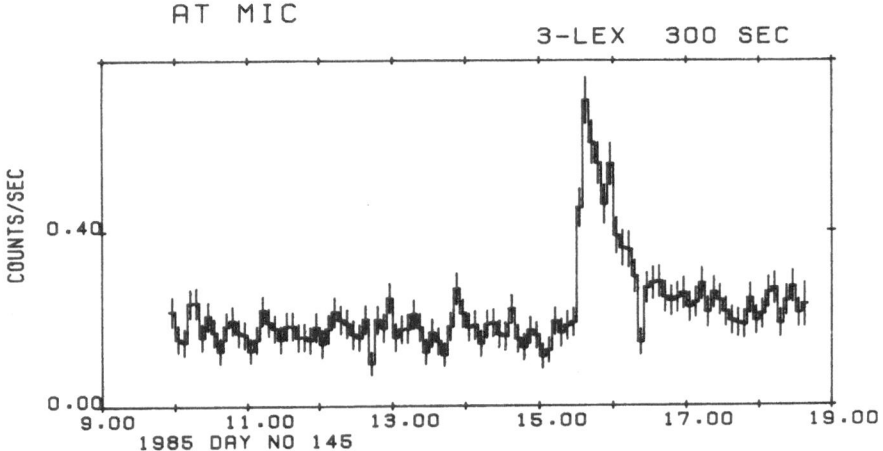

Fig. 6 - EXOSAT LE observation of a flare on the dMe star AT Mic
(reanalyzed data from the EXOSAT Archive; see also Nelson
et al. 1987).

ved A-type stars (A1V + A2Vm, both spectroscopic binaries). Since, as shown by EINSTEIN, A-type dwarfs are not strong X-ray emitters (if at all), the observed EXOSAT emission is mostly due to contamination of the EXOSAT detector by the ultra-violet radiation from the photosphere and the chromosphere of the star. The data in Fig. 7 show the occurrence of a flare - first reported by Droemer and Gibson (1985) - whose time behavior closely resembles that of flares observed on the Sun and dMe flare stars. It is interesting to speculate on the - presumably magnetic - activity of this star, which is thought not to possess a subphotospheric convective zone, and, hence, which should not have the high level of turbulent surface motions required to stress surface magnetic fields.

How can we model the observed events? If we take the solar analogy as a guide - line, we can consider a simple magnetic loop, whose configuration remains unchanged throughout the event. We further assume that energy is deposited impulsively close to the top of the loop, and that there is no energy input during the decay phase. This is what is believed to occur in *compact* flares on the Sun (Moore et al. 1980). The flare will decay through radiative and conductive losses which we can write as:

$$\tau_R = 3 \ K_B \ T \ / \ n \ P(T) \tag{5}$$

and

$$\tau_C = 4.8 \times 10^{-10} \ n \ L_C^2 \ T^{-5/2} \qquad sec \tag{6}$$

In the Eqs. above, T is the flare temperature, n the density, K_B is the Boltzmann constant, P(T) is the emissivity function for an optically thin plasma, and L_C is a characteristic length for the temperature gradient.

From Eqs. (5) and (6), and the temperature and emission measure derived from X-ray data, we can estimate the physical parameters of the flaring region (density, volume, characteristic length L_C ecc.). In order to do so, the observed decay time is equated to the radiative time, and the further assumption is made that conduc-tive and radiative times are approximately equal. This approach, as crude as it might be, gives reasonable numbers when applied to flares on dMe stars. Densities, temperatures and volumes derived in this way (T $\simeq$ 2-3X10^7 K, n $\simeq$ 10^{11}-10^{12} cm^{-3}, V $\simeq$ 10^{27}-10^{28} cm^3) are not very different from those of flares on the Sun. The total energy release, however, may be much larger, being typically of the order of 10^{31} - 10^{33} erg as compared to $\simeq$ 10^{29} - 10^{31} erg of compact flares on the Sun and $\simeq$ 10^{32} erg of large solar two-ribbon flares. Note, however, that the flare on Algol mentioned above (cf. Fig. 4) released at least $\simeq$ 10^{35} erg and involved a much larger volume ($\simeq$ 10^{31} cm^3) than typical solar flares. It also had a higher temperature (T $\simeq$ 6 X 10^7 K) and longer decay time ($\simeq$ 7 X 10^3 sec) than most flares on the Sun and on dMe flare stars (White et al 1986). Recently, Schmitt, Harnden and Fink (1987) have checked the reliability of the above mentioned approach by applying it to *solar flares* observed with the EINSTEIN Observatory by looking at the X-ray radiation scattered by the Sun-lit Earth. The parameters derived for these flares using the simple formalism above are in quite good agreement with those derived directly from spatially resolved solar observations.

Certainly, the physics involved in real flares is much more complex than the simple orders of magnitude estimates discussed above. A flaring loop is not an iso-lated system: it is rooted in the dense chromospheric and photospheric layers, and the flare time evolution will depend on the complex hydrodynamic phenomena resulting

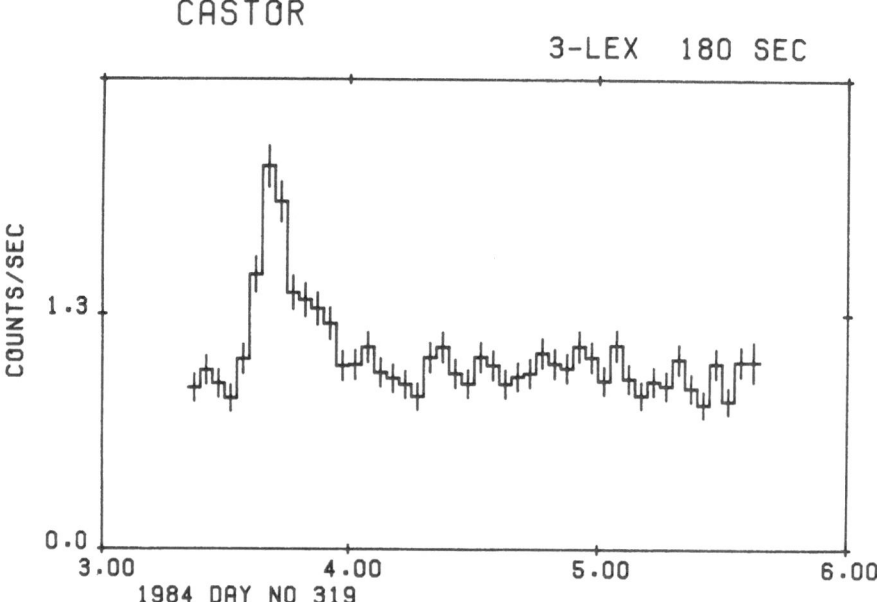

Fig. 7 - EXOSAT LE observation of a flare on the A-type star Castor.
The detected emission is mostly ultraviolet radiation from
the photosphere of the star (unpublished data from the
EXOSAT Archive; see also Droemer and Gibson 1985).

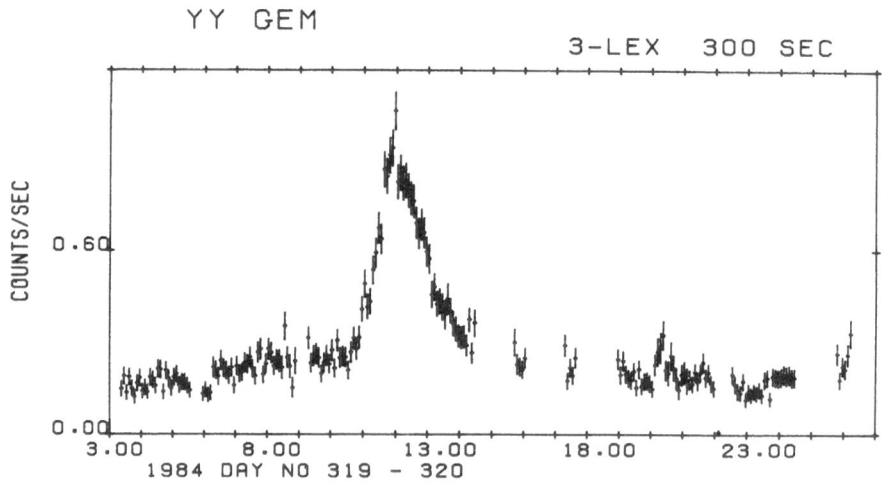

Fig. 8 - EXOSAT LE observation of the dMe eclipsing binary YY Gem.
A large flare dominates the light curve (unpublished data
from the EXOSAT Archive; see also Droemer and Gibson 1985).

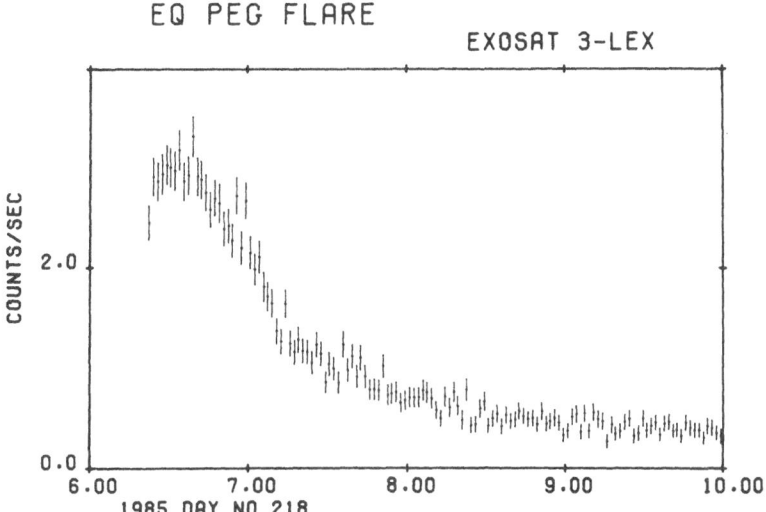

Fig. 9 - EXOSAT LE observation of a large long decay flare on
EQ Peg (from Pallavicini et al. 1986).

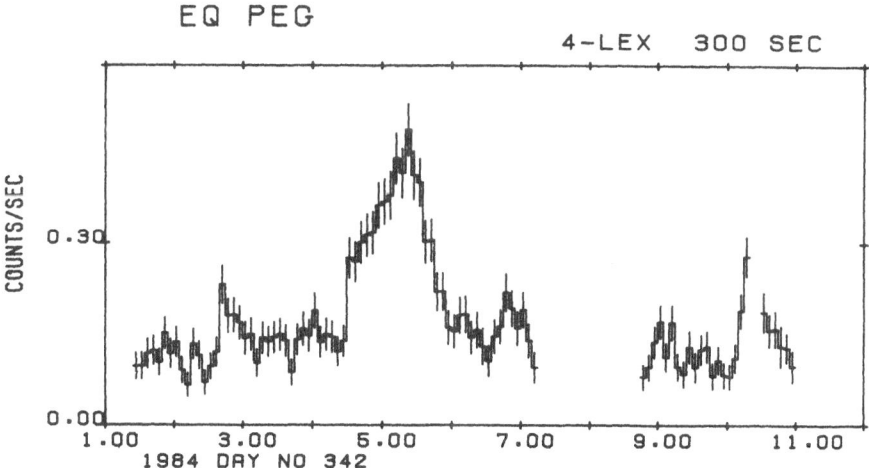

Fig. 10 - EXOSAT LE observation of an unusual flare on EQ Peg characterized by slow rise and more rapid decay (reanalyzed data
from the EXOSAT Archive; see also Haisch et al. 1987).

from this coupling. For instance, if energy is deposited at the loop top, it will be transferred to lower levels either by accelerated particles or by heat conduction. If the chromosphere receives more energy than it can radiate away, it will expand upwards (chromospheric evaporation) filling the loop with high density plasma. This in turn will profoundly affect the time evolution of the flare. Full hydrodynamic calculations of this type have been carried out for the Sun and have been successfully applied to flares observed from the Solar Maximum Mission (Pallavicini et al. 1983, Cheng et al. 1983, Peres et al. 1987). In principle the same type of modelling can be applied to flares on stars with the aim of getting better insights into the flare phenomenon. There are already a few groups, both in the States and in Europe, that are actively pursuing this question (e.g. Reale et al. 1987).

So far, we have assumed that the flare occurs in a closed magnetic structure which does not change throughout the event, as in solar compact flares. However, not all flares on the Sun behave in this way. Flares on the Sun are of two types (Pallavicini, Serio and Vaiana 1977, Priest 1981): a) compact, short-lived events, b) large, long-decay, two-ribbon events. These two classes of flares differ not only in their morphology and temporal evolution, but also in the basic physical mechanisms which operate throughout their lifetime. In particular, large two-ribbon flares on the Sun are believed to occur as a consequence of a disruptive phenomenon which suddenly opens a magnetic field structure (Kopp and Pneuman 1976, Kopp and Poletto 1984). The open field lines then relax back to a closed potential configuration, by gradually releasing energy as they reconnect during the decay phase of the flare. An interesting question is whether these events occur also on stars. I believe that the answer is affirmative. EXOSAT has recently observed a number of large, long-duration events, strongly reminiscent of solar two-ribbon flares. Examples are flares observed on EQ Peg and YY Gem (Figs. 8 and 9). Another example may be a long-decay flare on Prox Cen observed previously with EINSTEIN (Haisch et al. 1983). In order to test whether these flares are indeed the stellar counterpart of solar two-ribbon flares, Poletto, Pallavicini and Kopp (1986, 1987) have applied to some of them a reconnection model previously used successfully for solar two-ribbon flares. They have found that the model is capable of reproducing correctly the energy release rate and temporal evolution of the decay phase of the observed flares, thus supporting the correctness of the identification. Furthermore, this allows constraints to be put on the physical parameters of the flaring region (area extension, strength of the photospheric magnetic field, velocity of growth of post-flare loops etc.).

To conclude, the extensive coronal observations obtained with EINSTEIN and EXOSAT tend to support the view that a substantial analogy exists between phenomena observed in the solar corona and many (probably most) phenomena observed in the coronae of other late-type stars. This analogy allows us to use, for modelling stellar observations, the same basic techniques successfully developed for interpreting the much more detailed solar observations. However, given the widely different parameters occurring in late-type stars, we should also be careful not to exclude the possibility that stellar phenomena may exist, which do not have an exact solar counterpart. An example may be the extended coronal structures suggested by eclipse observations in RS CVn binaries. Another example, recently discovered with EXOSAT (Haisch et al. 1987, Pallavicini and Stella 1987), may be flares on dMe stars which are characterized by a gradual rise and a more rapid decay (Fig. 10). These apparently un-solar phenomena should be investigated carefully if we want to establish the extent to which the solar analogy can be correctly applied to other stars.

REFERENCES

Ambruster, C.W., Sciortino, S. and Golub, L.: 1987, submitted to Ap. J. Suppl.

Antiochos, S. and Noci, G.: 1986, Ap. J. 301, 440.

Athay, R.J. and White, O.R.: 1978, Ap. J. 226, 1135.

Brinkman, A.C., Gronenschild, E.H.B.M., Mewe, R., McHardy, I. and Pye, J.P.: 1985,
    Adv. Space Res. 5, No. 3, 65.

Bruner, E.C. Jr.: 1978, Ap. J. 226, 1140.

Bruner, E.C. Jr.: 1981, Ap. J. 247, 317.

Butler, C.J. and Rodonò, M.: 1986, Lecture Notes in Phys. 254, 329.

Butler, C.J., Rodonò, M., Foing, B.H. and Haisch, B.M.: 1986, Nature 321, 679.

Cheng, C.C., Oran, E.S., Doschek, G.A., Boris, J.P. and Mariska, J.T.: 1983, Ap. J.
    265, 1090.

Cowling, T.G.: 1981, Ann. Rev. Astron. Ap. 19, 115.

de Jager, C. et al.: 1986, Astron. Ap. 156, 95.

Droemer, D. and Gibson, D.M.: 1985, unpublished poster paper presented at 4th Cam-
    bridge Workshop on Cool Stars, Stellar Systems and the Sun, Santa Fé, NM, Octo-
    ber 1985 (see Abstract Booklet).

Doyle, J.G. and Butler, C.J.: 1985, Nature 313, 378.

Felli, M., Massi, M., Palagi, F., Pallavicini, R. and Tofani, G.: 1987, in prepara-
    tion.

Giampapa, M.S., Golub, L., Peres, G., Serio, S. and Vaiana, G.S.: 1985, Ap. J. 289,
    203.

Gilman, P.A. and De Luca, E.E.: 1986, Lecture Notes in Phys. 254, 163.

Haisch, B.M., Butler, C.J., Doyle, J.G. and Rodonò, M.: 1987, Astron. Ap., in press.

Haisch, B.M., Linsky, J.L., Bornmann, P.L., Stencel, R.E., Antiochos, S.K., Golub, L.
    and Vaiana, G.S.: 1983, Ap.J. 267, 280.

Kopp, R.A. and Pneuman, G.W.: 1976, Solar Phys. 50, 85.

Kopp, R.A. and Poletto, G.: 1984, Solar Phys. 93, 351.

Kuperus, M., Ionson, J.A. and Spicer, D.S.: 1981, Ann. Rev. Astron. Ap. 19, 7.

Kundu, M.R., Pallavicini, R., Jackson, P. and White, S.M.: 1987, in preparation.

Landini, M., Monsignori-Fossi, B.C. and Pallavicini, R.: 1985, Space Science Rev.
    40, 43.

Landini, M., Monsignori-Fossi, B.C., Pallavicini, R. and Piro, L.: 1986, Astron. Ap.
    157, 217.

Landini, M., Monsignori-Fossi, B.C., Paresce, F. and Stern, R.A.: 1985, Ap. J. 289,
    709.

Lin, R.P., Schwartz, R.A., Kane, S.R., Pelling, R.M. and Hurley, K.C.: 1984, Ap. J.
    283, 421.

Lucy, L.B. and White, R.L.: 1980, Ap.J. 241, 300.

Majer, P., Schmitt, J.H.M.M., Golub, L., Harnden, F.R. Jr. and Rosner, R.: 1986, Ap.
    J. 300, 360.

Mewe, R.: 1979, Space Science Rev. 24, 101.

Mewe, R., Schrijver, C.J., Lemen, J.R. and Bentley, R.D.: 1986, Adv. Space Res., in
    press.

Moore, R. et al.: 1980, in Solar Flares (P.A. Sturrock ed.), p. 341.

Mutel, R.L., Lestrade, J.F., Preston, R.A. and Phillips, R.B.: 1985, Ap. J. 254, 641.

Nelson, G.J., Page, A.A., Slee, O.B. and Denby, B.: 1987, M.N.R.A.S., in press.

Pallavicini, R., Golub, L., Rosner, R., Vaiana, G.S., Ayres, T. and Linsky, J.L.:
    1981, Ap. J. 248, 279.

Pallavicini, R., Kundu, M.R. and Jackson, P.D.: 1986, Lecture Notes in Phys. 254, 225.

Pallavicini, R., Monsignori-Fossi, B.C., Landini, M. and Schmitt, J.H.M.M.: 1987,
    Astron. Ap., in press.

Pallavicini, R., Peres, G., Serio, S., Vaiana, G.S., Acton, L., Leibacher, J. and Rosner, R.: 1983, Ap. J. 270, 270.

Pallavicini, R., Serio, S. and Vaiana, G.S.: 1977, Ap. J. 216, 108.

Pallavicini, R. and Stella, L.: 1987, in preparation.

Parker, E.N.: 1983, Ap. J. 264, 642.

Peres, G., Reale, F., Serio, S. and Pallavicini, R.: 1987, Ap. J. 312, 895.

Poletto, G., Pallavicini, R. and Kopp, R.A.: 1986, Adv. Space Res., in press.

Poletto, G., Pallavicini, R. and Kopp, R.A.: 1987, in preparation.

Priest, E.R.: 1981, in Solar Flare Magnetohydrodynamics (E.R. Priest ed.), p.1.

Reale, F. et al.: 1987, in preparation.

Rosner, R., Tucker, W.H. and Vaiana, G.S.: 1978, Ap. J. 220, 643.

Rosner, R., Golub, L. and Vaiana, G.S.: 1985, Ann. Rev. Astron. Ap. 16, 393.

Schmitt, J.H.M.M.: 1984, X-Ray Astronomy '84 (M. Oda and R. Giacconi eds.), p. 17.

Schmitt, J.H.M.M.: 1987, in preparation.

Schmitt, J.H.M.M., Harnden, F.R. Jr. and Fink, H.: 1987, Ap. J., in press.

Schmitt, J.H.M.M., Golub, L., Harnden, F.R. Jr., Maxson, C.W., Rosner, R. and Vaiana, G.S.: 1985a, Ap. J. 290, 307.

Schmitt, J.H.M.M., Harnden F.R. Jr., Peres, G., Rosner, R. and Serio, S.: 1985b, Ap. J. 288, 751.

Schmitt, J.H.M.M., Pallavicini, R., Monsignori-Fossi, B.C. and Harnden, F.R.: 1987, Astron. Ap., in press.

Skumanich, A.: 1985, Aust. J. Phys. 38, 971.

Stern, R.A., Antiochos, S.K. and Harnden, F.R. Jr.: 1986, Ap. J. 305, 417.

Swank, J.H., White, N.E., Holt, S.S. and Becker, R.H.: 1981, Ap. J. 246, 208.

Uchida, Y. and Sakurai, T.: 1983, in Activity in Red-dwarf Stars (P.B. Byrne and M. Rodonò eds.) p. 629.

Vaiana, G.S. and Rosner, R.: 1978, Ann. Rev. Astron. Ap. 16, 393.

Vaiana, G.S. et al.: 1981, Ap. J. 245, 163.

Walter, F.M., Gibson, D.M. and Basri, G.S.: 1983, Ap. J. 267, 665.

White, N.E.: 1987, private communication.

White, N.E., Culhane, J.L., Parmar, A.N., Kellett, B.J., Kahn, S., van den Oord, G.H.J. and Kuipers, J.: 1986, Ap. J. 301, 262.

Whitehouse, D.R.: 1985, Astron. Ap. 145, 449.

STELLAR vs. SOLAR ACTIVITY:

THE CASE OF PRE-MAIN SEQUENCE STARS

Thierry Montmerle

Service d'Astrophysique
Centre d'Etudes Nucléaires de Saclay
91191 Gif-sur-Yvette, France

ABSTRACT

X-rays have proved a powerful tool in discovering, or monitoring, the activity of many kinds of stars. Among these, the T Tauri stars and other pre-main sequence objects are particularly interesting in that, based on X-rays, the activity seems to be of solar type, albeit on a scale 10^3 to 10^6 times higher in intensity, mainly in the form of gigantic flares. Also, in the radio range, follow-up observations have allowed to discover a few flaring PMS stars. Optical results (multiband photometry) have recently confirmed the solar nature of this activity to some extent, by revealing periodic variations attributed to large starspots. In addition, from these data and other rotation indicators, one finds a fair correlation between rotation and activity, as can be expected from dynamo-induced phenomena already known on the Sun.

However, this seems to be only part of the story. Already in the visible, such traditional activity indicators as Hα point to the existence of large sources of energy in forms other than magnetic. In the microwave range, exotic phenomena like jets or bipolar flows appear to be widespread. In the radio cm range, the emission, when detected, is non-solar in most cases, and intriguing examples from recent VLA observations are given. The reasons for this may lie in the specific nature of PMS objects, i.e., very young and perhaps still surrounded by circumstellar material (e.g., accretion disks), drawn from their parent molecular cloud.

1. THE SOLAR-STELLAR CONNECTION

That some stars share at least some of the properties of the active Sun has been kown for a long time in the optical domain. For instance, although individual light curves are usually not available, strong variability (several magnitudes in time scales at hours, or even minutes) has been interpreted as flares in dMe, UV Ceti, or T Tauri stars ; on other stars, periodic variability has been interpreted as being due to large starspots, like in RS CVn or BY Dra close binaries.

But all these stars were "abnormal" in some way, and it was not until the advent of, first, solar X-ray astronomy (Skylab, 1973), and second, stellar X-ray astronomy (Einstein, 1978-1981), that it became clear that "solar-type" activity was widespread within the Herzsprung-Russell diagram, and, most remarkably, among ordinary stars. One major reason for this: the contrast between quiescent and active states is, in general, much higher in X-rays than at other wavelengths, and stellar X-rays can be seen from large distances (several kpc). Since the detected < keV X-rays correspond to temperatures $\approx 10^6$-10^7 K, they are however by themselves insufficient to characterize stellar outer atmospheric structures other than flares or coronae (for a review, see e.g., Rosner, Golub, and Vaiana, 1985). Hence the equally crucial role played by IUE to characterize chromospheres (T $\lesssim$ 10^4 K), along with continuing photospheric studies from the ground.

It is now widely recognized that a "solar-stellar connection" exists. This is meant to say that, by using a number of quantitative tracers to measure the activity of stars, there exists only a simple scale factor between the Sun and the various stars studied ; this scale factor may reach several orders of magnitude. In turn, this scale factor may be compared with other stellar parameters to help understand the origin of stellar activity in general (e.g., Pallavicini, 1985).

This "solar-stellar connection", however, appears restricted mostly, if not entirely, to late-type stars, i.e., of spectral type later than $\approx$ early F. For instance, flares or coronae (at least in the solar sense) do not exist in O and B stars ; their X-ray "efficiency" L_X/L_{bol} is much smaller than for late-type stars ($\approx 10^{-7}$ vs. $\gtrsim 10^{-4}$), and does not depend on the spectral type. The transition appears around type A, where the activity seems minimal.

To date, the conventional wisdom is that convection and rotation play a central role in generating the activity: (i) the stars cease to have an outer convection zone at mid-F (there is an increasingly large convective core for earlier types) ; (ii) fast rotators (young stars, or older stars rotating rapidly because of tidal synchronism with a close companion) are more active than slower ones. This supports the dynamo picture of stellar activity: through the dynamo mechanism (e.g., the "α-ω dynamo", see Gilman 1983), convection and (differential) rotation generate surface magnetic fields ; in turn, because of the surface movements also linked with convection (see, for instance, the Sun), the breaking and subsequent reconnection of field lines create energy in the form of flares, coronae, etc...

However, one is still looking for a more satisfactory situation than the above qualitative description. Indeed, in spite of all the observational and theoretical material at hand, the choice of the relevant tracers and physical parameters is still debated (e.g., Basri 1987a, b). For instance, to characterize activity, should one use total fluxes (in Mg II, Ca II, X-rays, etc...) ? Or fluxes per unit area, or fluxes normalized to the bolometric flux, or "excess fluxes" ? To characterize rotation, should one use v sini ? Or the period ? Or add a convection ingredient, in the form of the "Rossby number" R_o = Period/τ_c (where τ_c is the convective turnover time), even if this means mixing an observationally measured value (Period) and a theoretical, model-dependent, result (τ_c) ?

To help answer these questions, it is useful to broaden the star sample as much as possible, looking for wide ranges in the various parameters (up to orders of magnitude): rotation velocity, depth of convection zones, spectral type, binarity, etc. In this context, pre-main sequence (PMS) stars are very interesting: the bulk of the PMS stars known belong to spectral types later than F5, but are more deeply (even fully) convective, more luminous, and rotate faster than their main-sequence counterparts. A potential complication, however, is that the purely stellar picture may be blurred by the influence of the environment (molecular clouds, circumstellar material, etc.).

The present review will therefore concentrate on "Young Stellar Objects" (YSOs). This designation, although rather ill-defined, is used more and more frequently, in part to emphasize the fact that the classical T Tauri stars (see, e.g., Bertout, 1984), or even PMS stars, form only a subset of YSOs, which are now being discovered in increasingly large numbers at essentially all wavelengths, mainly in the radio, infrared, and X-rays, in addition to the traditional optical. We will put some emphasis on the nearly ρ Oph cloud star formation region, ≈ 160 pc away, which is a particularly good example of a multi-wavelength study over a large area (several square degrees).

We first look for solar-type activity in the sense outlined above ; the results tend to support the general picture (§ 2). However, there is also evidence for <u>non-solar activity</u> (co)existing with it, in forms often "exotic" (jets, bipolar flows, etc...) (§ 3). Finally, we look to what extent it is possible to reconcile some apparently conflicting results by using our knowledge of the environment(s) of YSOs (§ 4), before making some concluding remarks (§ 5).

2. EVIDENCE FOR SOLAR-TYPE ACTIVITY

2.1 Flares

Several regions of star formation, as well as individual T Tauri stars, have been observed in X-rays by the <u>Einstein</u> Observatory and, to a lesser extent, by EXOSAT. These observations have provided the best evidence to date that YSOs are frequently flaring, and that the flares are of a solar nature (see reviews by Feigelson 1984, 1987). In fact, the largest existing body of evidence for stellar X-ray flares comes essentially from the observation of such objects.

The early individual cases of two classical T Tauri stars, DG Tau and AS 205 (Feigelson and De Campli 1981 ; Walter and Kuhi 1984) have been supplemented by the results of Montmerle et al. (1983, 1984), who obtained repeated exposures of the ρ Oph cloud with the IPC instrument, covering ≈ 2° x 2°, i.e. overlapping the darkest parts of the cloud. About 50 sources were discovered, characterized by a generalized variability (the "ρ Oph Christmas tree"), involving factors ≈ 2-20. Almost all these "ROX" sources are YSOs, about 10 being previously known classical T Tauri stars. The remainder were suspected, and later confirmed (see below, § 2.2) to be pre-main sequence objects, many of them being in fact faint, weak T Tauri stars. The demonstration that the observed variability must be interpreted in terms of solar-type flares rests essentially on three arguments, as follows:

(i) In one particular instance, the source ROX-20 underwent a strong event, the decay of which could be followed on a timescale of ≈ 2 hrs (fig. 1a) ; this was interpreted as the radiative cooling phase of a very strong flare (perhaps the strongest stellar flare ever recorded): $L_{x,max} \approx 10^{32}$ erg.s^{-1}, $T \approx 1$ keV (deduced from a reasonably good bremsstrahlung continuous spectrum), density $n \approx 10^{10}$ cm^{-3}, size $\ell \approx 10^{12}$ cm, i.e., 3 $R_{\odot}$ or ≈ 1 stellar radius R_{*}. This event, therefore, has a typical solar flare density and temperature ; its high luminosity is essentially

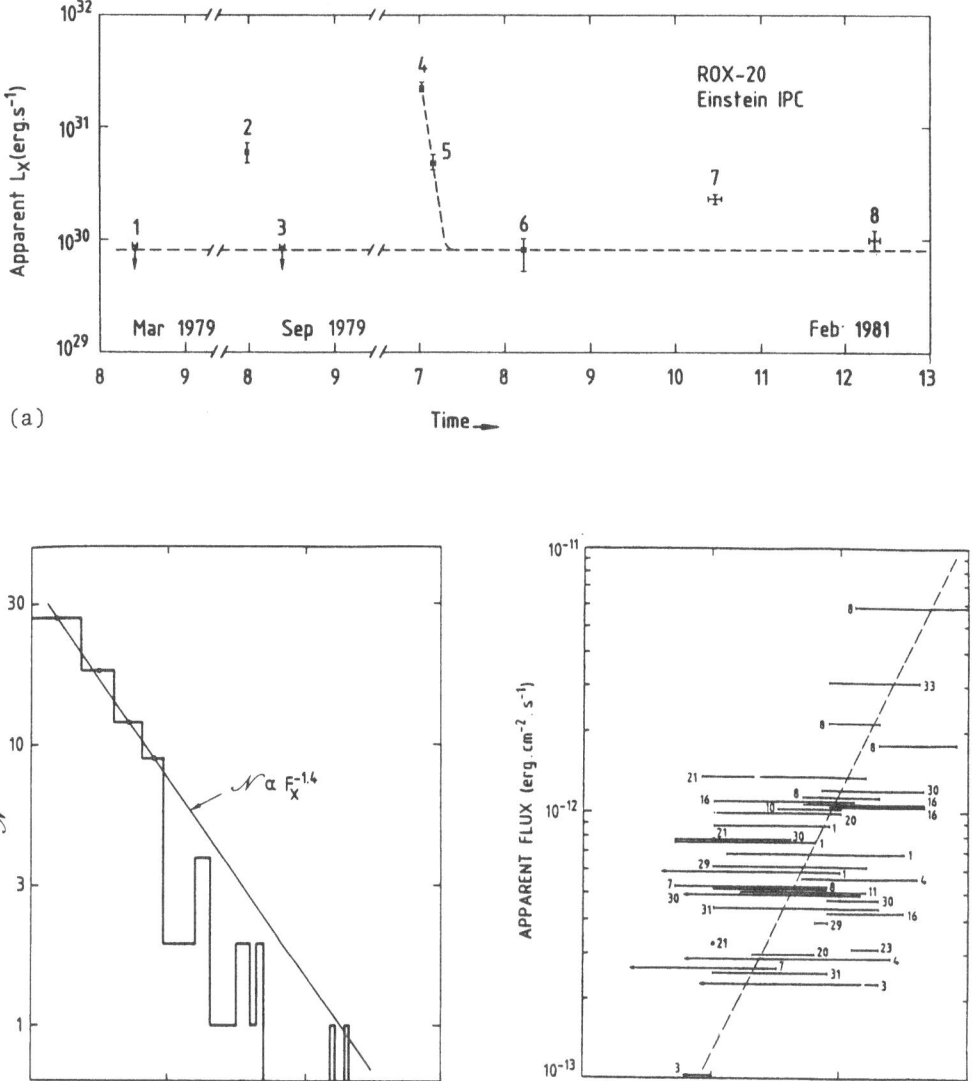

Fig. 1. (a) *Light curve of the Rho Oph X-ray source ROX-20, obtained by repeated exposures of the Einstein Observatory. The event corresponding to observations 4, 5, and 6 is interpreted as the cooling phase of a very strong flare. (b) Cumulative flux distribution of the X-ray flares of the ROX sources. The same distribution (with a larger number of events) holds also for the Sun (see Montmerle et al. 1983). (c) Flux distribution as a function of the temperature. (Numbers refer to the ROX sources; the dotted line is only a guide to the eye. Montmerle et al. 1984.)*

due to its enormous size. As Table 1 shows, this is just an extreme case, following the same trend as other stellar flares in general.

(ii) Given that all X-ray detections are very similar, differing only in their intensity (in line with the above remark, the X-ray spectra, when available, do not differ much), one can put them all into a "log N - log S"-like histogram. As fig. 1b shows, one obtains roughly a power-law, of index -1.4, i.e., the same as that of the similar histogram obtained on the Sun.

(iii) Another typically solar feature, displayed on Fig. 1c, although less clearcut, is that the X-ray temperature tends to increase with the flare luminosity.

The solar nature of this activity has been confirmed in a few cases in the radio (GHz) range. Indeed, in the course of a follow-up survey of the ρ Oph cloud with the NRAO Very Large Array to look for radio counterparts to the ROX sources (see André et al., 1987), rapid variability in the form of a flux increase of a factor < 2 over ≈ 2 hrs was found in DoAr21, one of the strongest X-ray sources of the cloud (Fig. 2a, Feigelson and Montmerle, 1985). The maximum flux observed can be interpreted as gyrosynchrotron emission of ≈ MeV electrons in a magnetic field of about 100 G over a loop of size ≈ several stellar radii. This size is so large that one needs in fact to resort to inhomogeneous models of synchrotron emission (Klein and Trottet, 1984): as shown on Fig. 2b, where a dipolar field has been used, different frequencies are associated with different regions of the loop (see André, 1987 ; also Klein and Chiuderi-Drago, 1987). Fig. 2c displays the long timescale behaviour of DoAr21, which shows that variability is always present, but that strong fluxes are comparatively rare. A similar flare-like behaviour has been found in another source, ROX-31 (Stine et al. 1987, in preparation).

To date, these are the only known cases of rapid radio variability in YSOs, but large variabilities have also been observed between long time intervals (months or years) in other objects (V410 Tau, HP Tau/G2 and /G3, Cohen and Bieging, 1986), making the flare interpretation likely in these cases as well, especially in view of the long-term variability of sources like DoAr21 or ROX-31.

Table 1. Properties of stellar flares (see Montmerle et al. 1983)

Star		L_x(max) (10^{30} ergs s^{-1})	T (10^7 K)	Size (10^9 cm)	Density (10^{10} cm^{-3})
Sun	compact flares	$\sim 10^{-4}$	1–2	$\sim 10^{-3}$	10–100
	flares	$\sim 10^{-3}$	1–2	$\gtrsim 1$	10
	class 3 flares	$> 10^{-3}$	1–2	> 10	1–10
dMe stars	AD Leo	1.6	(3)	(~ 3)	(10)
	YZ CMi[a]	8	(1)	...	...
	AT Mic	16	3	(~ 7)	(10)
	Prox. Cen[b]	1.8	(1)	...	...
	UV Cet	2.0	(~ 1)	...	...
T Tau and related stars	ROX[c] sources	~ 5	2	(10–30)	(10–20)
	DG Tau	$\gtrsim 20$	(1)	[50]	[10]
	ROX 20	~ 100	1–2	~ 100	3–6

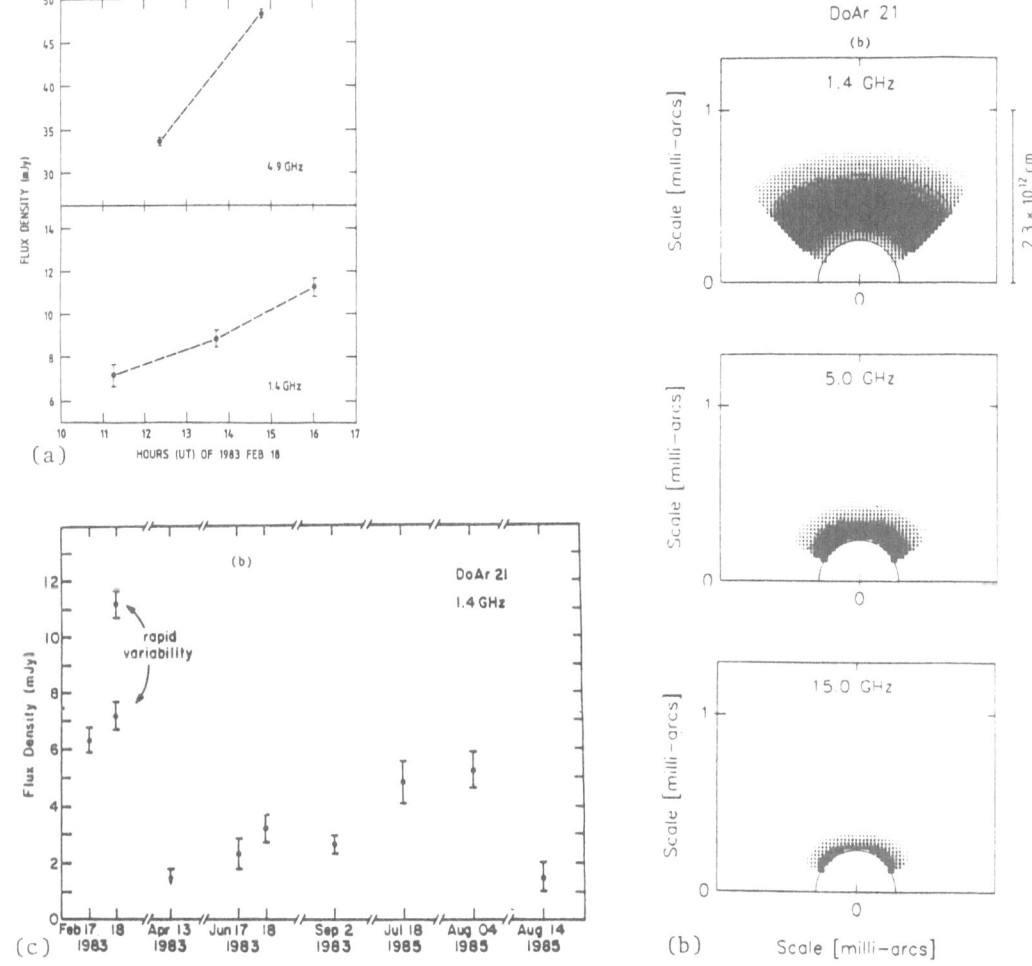

Fig. 2. *Radio observations of DoAr21(= ROX8). (a) Rapid variability (Feigelson and Montmerle 1985); (b) interpretation in terms of inhomogeneous gyrosynchrotron emission associated with a flare event in a dipolar magnetic loop (see André 1987, using the model of Klein and Trottet 1984); (c) long timescale behavior (André 1987).*

Note, however, in contrast with the X-ray situation, that strong radio flares are rare, and concern only a small number of objects.

2.2. Starspots

Another well-known piece of evidence regarding solar-like activity is the presence of starspots. The evidence is always indirect, as no images are seen ; it relies on repeated observations either of near-IR and optical photometry or of a photospheric tracer like the CaII H and K lines.

We will briefly summarize the recent results of the first method, obtained on a sample of T Tauri stars (Bouvier et al., 1986, Bouvier and Appenzeller, in preparation, and Bouvier 1987), which were confirmed for the targets in common by the second method (Hartmann et al., 1987). The sample, in particular, includes a number of ROX sources.

The use of UBVRI photometry allows to derive a temperature T (by fitting a black body), hence its variation as a function of time. A model may then be constructed, in which the smaller fluxes are assumed to be associated with circular starspots of temperature differing by an amount ΔT from the photospheric temperature T_{eff}, at a latitude β, and covering f % of the total stellar surface (for details, see Bouvier 1987a, 1987b). While this modelization may be considered as simplistic, it involves a minimum number of parameters and gives a reasonably good fit to the data (Fig. 3a). In almost all cases, the starspot interpretation must be basically correct, since ΔT is < 0 ; in one case, however (DF Tau), ΔT is > 0: the "spot" is hotter than the surrounding area, and at a high latitude. We shall come back to this point below ($\S$ 4.1).

If one measures the activity of T Tauri stars in the sample by the filling factor f of their starspots, then these stars appear here also as very active, on the same level as RS CVn stars, in any case much more active than their main-sequence counterparts (Fig. 3b).

In addition to giving a measure of activity, starspots give also at the same time the rotation period, which is of course precious in the context of the rotation-convection vs. activity connection.

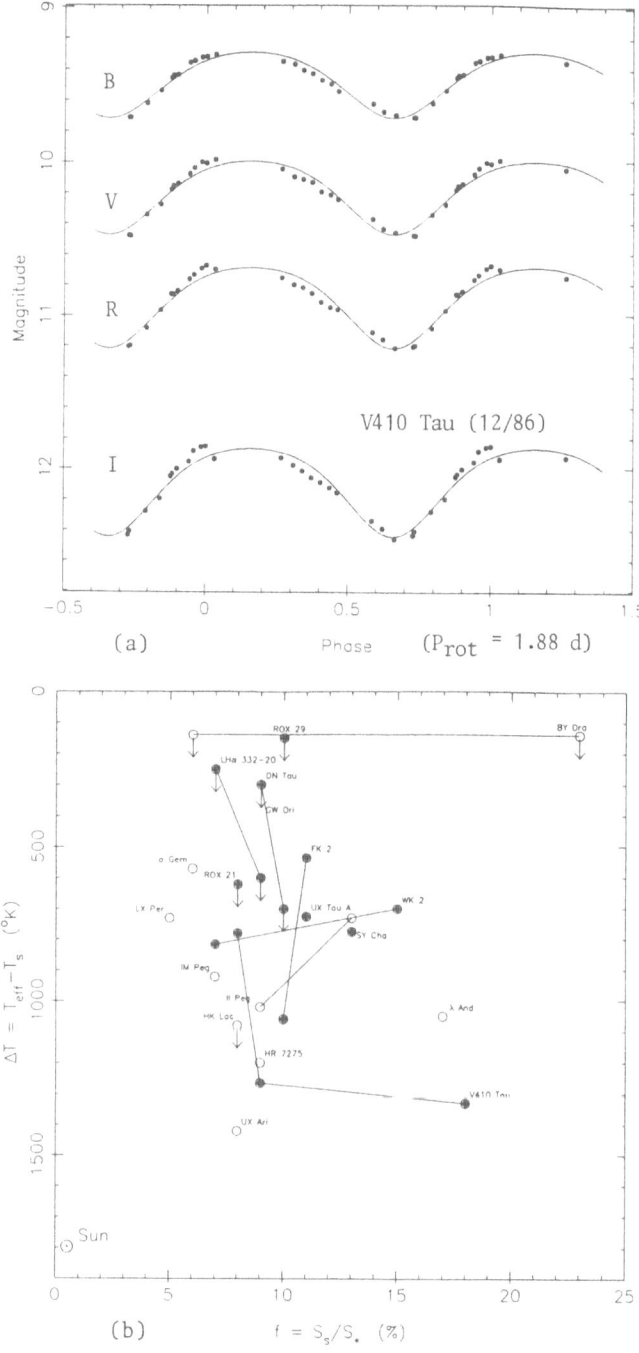

Fig. 3. (a) BVRI photometry of the active T Tauri star V410 Tau. The light curve is interpreted in terms of a high-latitude starspot, covering f = 17% of the stellar surface, and cooler by ΔT = 1425 K. (b) ΔT vs. f diagram for T Tauri stars (dark spots), and for BY Dra and RS CVn binary systems (a and b from Bouvier 1987a).

2.3. Activity vs. rotation and convection

The T Tauri stars studied by Bouvier (1987a, 1987b) all have known X-ray fluxes F_x. Taking these as an activity tracer, one may then plot F_x against the rotation period P, or against the Rossby number $R_0 = P/\tau_c$ (see above, § 1). As shown on Fig. 4a, where the results on T Tauri stars have been supplemented by similar data on a sample of stars of similar spectral types (main-sequence dwarfs and RS CVn binaries), there is definitely a trend for faster rotators to be more active. However, at a given period, T Tauri stars are -as expected- more active than dwarfs. We note that this is at variance with the "shell dynamo" explanation of activity (i.e., dynamo generated at the <u>bottom</u> of the convective zone), since T Tauri stars are fully convective, contrary to dwarfs. The RS CVn stars are quite scattered, but, at the same rotation period, they tie in rather nicely with T Tauri stars.

Fig. 4b shows the same points, but using the Rossby number R_0. The trend is still there, but in a more confused fashion. In particular, the T Tauri stars appear much more scattered: they span $\approx$ 2 orders of magnitude in F_x vs. less than

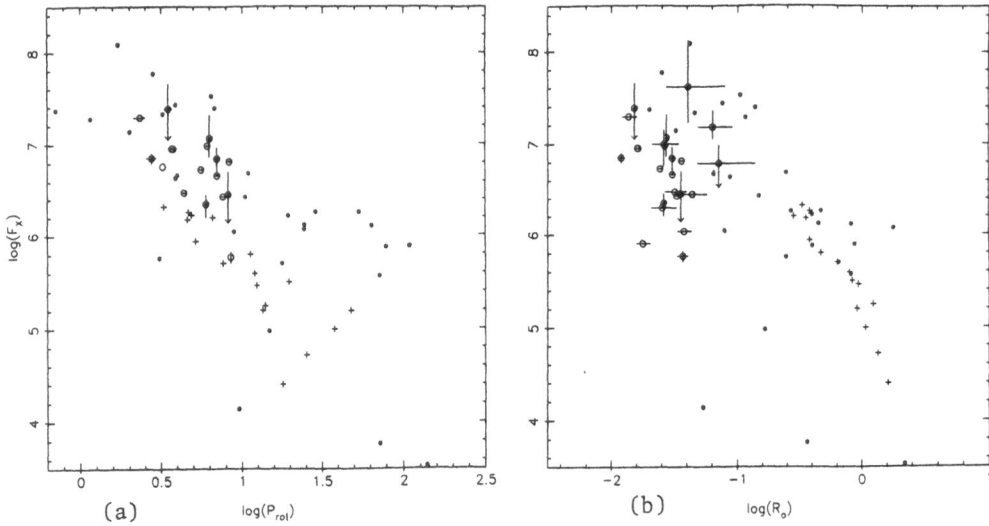

Fig. 4. Surface X-ray flux (in erg.cm^{-2}.s^{-1}) vs. P_{rot} (in days) (a), and vs. the Rossby number $R_0 = P_{rot}/\tau_c$ (b). Open circles: T Tauri stars, including some ROX sources; crosses: dwarfs; filled circles: RS CVn binaries (Bouvier 1987a,b).

one in R_o, and yet no correlation is apparent. This example illustrates how difficult it is to date to derive physically meaningful correlations.

But the main result is that, altogether, apart from a scale factor, the T Tauri stars follow the same trend as dwarfs and RS CVn stars in X-rays, confirming qualitatively the solar nature of their activity in this wavelength range.

3. EVIDENCE FOR NON-SOLAR ACTIVITY

3.1. Excess flux from several tracers

As stressed in the Introduction (§ 1), other activity tracers exist: CaII, MgII, Hα, etc... One can therefore draw the same kind of graphs than in the preceding section (Bouvier 1987a,b). The results are markedly different: there is a clear excess of those tracers in the case of T Tauri stars over the general trend. Figs. 5a and 5b show the case of F(Hα) vs. P or R_o. The situation is identical for CaII and MgII, indicating that the corresponding excess is located close to the surface of the star (photosphere and chromosphere). This is also the case for Hα,

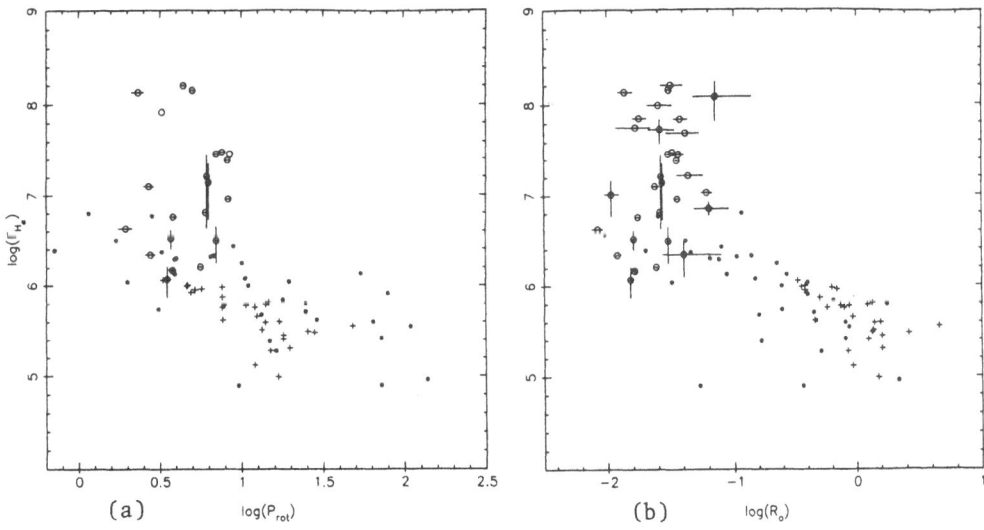

Fig. 5. Same as Fig. 4, for the Hα flux. There is a clear excess of Hα flux for some T Tauri stars, over the general trend.

if one brings in the X-ray result (§ 2.3) which relates to more extended outer regions: contrary to the early tendency, advocated by Walter and Kuhi (1981), that T Tauri stars have "smothered coronae", the X-ray emitting T Tauri stars have "normal" Hα (i.e., along the same trend as RS CVn stars or dwarfs), but some of them clearly have 10 to 100 times more Hα flux than others , while having about the same X-ray flux. In those cases, there is evidence for substantial, or even dominant, "non-magnetic" activity. Its nature, as traced mainly by Hα, is unclear ; it probably indicates, when present, the existence of a large amount of ionized gas, possibly in addition to the "normal" solar-like active regions over the stellar surface (of course, the presence of this gas cannot be explained by photoionization, since the stellar temperatures are too small). We will return to this point in § 4.1.

3.2. "Exotic" phenomena

Mass outflows are widespread among YSOs, and may take a variety of forms, from (spherical) winds, to narrow jets or very extended bipolar flows seen in CO.

Winds have been found in a variety of T Tauri stars from P Cygni profiles (in the optical or in the UV)(e.g. Bertout, 1984) or related Herbig Ae/Be stars (Catala et al., 1986). The velocities range from ≈ 150 km.s^{-1} to < 300 km.s^{-1}, i.e., are generally smaller than the solar wind. Mass-loss rates, on the other hand, have been derived in the past mostly from near-IR or radio measurements, with results sometimes in excess of $\approx 10^{-7}$ M$_\odot$ yr^{-1}. Theoretical considerations, (e.g. De Campli, 1981), as well as new data and interpretations (see below, for instance, the case of radio measurements) give a maximum value of a few 10^{-8} M$_\odot$ yr^{-1} at most, and more generally $\dot{M} < 10^{-8}$ M$_\odot$ yr^{-1}. Although smaller than the earliest estimates, this value remains comparatively high, $\approx 10^6$ times the values of $\dot{M}$ for the solar wind.

The geometry of the mass loss is, in general, not readily accessible. If the wind is ionized, a typical radio flux (see, e.g., Panagia and Felli, 1975) of a few mJy in the GHz range give sizes of order $\approx 10^{14}$ cm ; i.e. $\approx 10^3$ R$_*$. Such sizes cannot be resolved in general, except by the VLA in its most extended configuration, and for the closest sources (see, e.g., André, 1987). In an increasingly large number of cases, the mass loss takes the form of highly extended collimated jets (e.g. Mundt et al., 1984), of sizes up to $\approx 10^{15}-10^{16}$ cm, visiblé mostly in $H\alpha$.

The most extreme cases of mass loss are associated with cool bipolar flows (e.g., Lada, 1985 ; see also Casoli, 1987, Guilloteau, 1987), expanding over $\approx 10^{18}$ cm or more, with typical velocities again in the range $\approx 200-300$ $km.s^{-1}$. In this case, the associated mass loss is huge, reaching $\approx 10^{-4}-10^{-3}$ M_Θ yr^{-1} (but see discussion in Cabrit and Bertout, 1986, 1987), pointing to rather short-lived phenomena. The once enigmatic Herbig-Haro objects appear closely associated with jets or bipolar flows (e.g., Schwartz, 1983).

The physical mechanism of the mass loss is unknown, but must be very efficient since it sometimes reaches a few percent of the bolometric luminosity of the exciting object (see discussion in Lada, 1985), just as in the case of the massive Wolf-Rayet stars. For "ordinary" winds associated with T Tauri stars, however, heating (i.e., acceleration) by Alfvén waves has been advocated (e.g., Lago, 1982, Hartmann et al., 1982), but it should be stressed that the mass loss mechanism for the Sun is not known either...

In other YSOs, there is evidence for mass inflow, i.e., accretion. It may be spherical (see the example of T Tauri(S) itself, Bertout, 1983, André, 1987), or may compete with mass loss and generate Rayleigh-Taylor instabilities (YY Ori stars, see Mundt 1981). The accretion can also take place in the form of a disk, as will be discussed below, § 4.1.

3.3. Radio emission

The radio emission from the Sun has been known for a long time (see e.g., the Nançay radioheliograph), but, because of sensitivity problems, the field of stellar radioastronomy has initially developed rather slowly. The first star detected in the GHz range, has been, pointedly, a flare star (1963) a few pc distant ; the first T Tauri stars were detected in 1974 (T Tau, Lk Hα 101) ; but it was not until 1981, with the completion of the Very Large Array (allowing to reach ≈ 0.1

mJy in 8 hrs) that the field began to grow rapidly (e.g., Hjellming and Gibson, 1985).

The mechanism for radio emission may be either thermal (free-free emission from an ionized gas: winds, accretion, HII regions), or non-thermal (e.g., gyrosynchrotron from mildly relativistic electrons, as in the case of the DoAr21 flare, § 2.1 above) ; for reviews, see e.g., Dulk (1985) and André (1987). Recent observations have yielded suprising results for YSOs, which we now summarize.

To date, essentially two surveys of YSOs with the VLA exist. The first one is a survey of 41 classical T Tauri stars (Cohen and Bieging, 1986, and refs. therein); the other is the radio follow-up of the ρ Oph Einstein survey (André et al., 1987), which covers over 100 YSOs detected in the same area at X-ray, optical or IR wavelengths. At a sensitivity of > 1 mJy, the detection rate is low, on the order of 10 % in both surveys: 3 objects in the first survey, 12 in the second. But the most intriguing results come from the spectra (given by observations at two or three frequencies: 1.4, 5, or 15 GHz), of which all kinds are found - by contrast with the fair uniformity of the X-ray properties:

(i) "standard" winds (radio spectrum $\alpha \ \nu^{+0.6}$) are very rare, giving upper limits to the mass loss rates $< 10^{-8} \ M_\Theta \ yr^{-1}$;

(ii) only a few flaring objects have been found (see above, § 2.1), always detected on the basis of time variability - spectra alone are not sufficient (see below the case of WL5) ;

(iii) the present data have comparatively large error bars, hence allow several interpretations, including collimated winds (Reynolds, 1986) as shown on Fig. 6.

The situation holding in the ρ Oph survey (André et al., 1987) seems to be the paradoxical evidence for simultaneous thermal and non-thermal features (Fig. 6): source WL5 has a flarelike spectrum, but has remained constant over several different observations ; source VS14 is consistent with a thermal accretion flow, but has been found to be circularly polarized (at the 3σ level) , hence perhaps non-thermal ; source S1 has a resolved core + halo structure (Fig. 7), with a non-

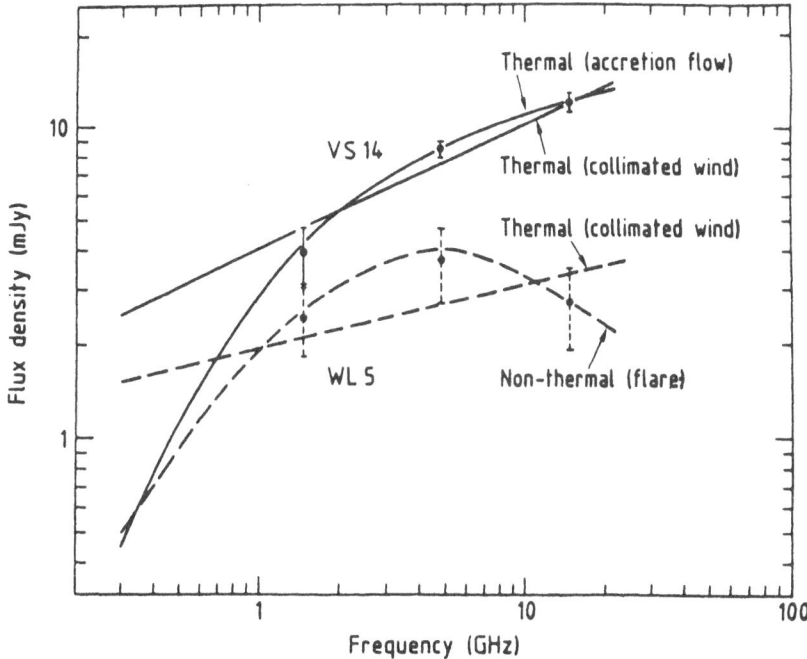

Fig. 6. _Possible interpretations of the VLA spectra of sources VS14 and WL5 in_
the ρ Oph cloud (André et al. 1987, André 1987).

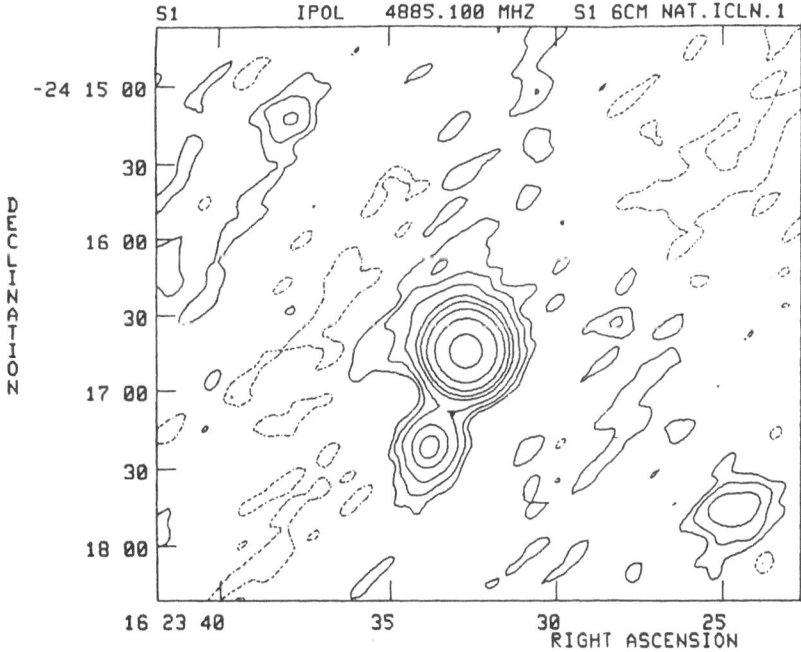

Fig. 7. _Map at 6 cm of source S1 in the ρ Oph cloud, obtained with the VLA in C/D_
configuration in February 1987. The synthesized beam size is 17″4 x 17″1, i.e.,~
4 x 10^{13} cm at the distance of the cloud. The main source has a clear core + halo
structure; the neighboring sub-mJy source was unknown and is as yet unidentified
(André et al. 1987, in preparation).

variable flat spectrum, suggesting a compact HII region, but is circularly polarized at the 10σ level (≈ 7 %). Obviously, there are several classes of radio-emit-
ting objects – perhaps as many as there are sources. A case by case study is in
progress (André et al., 1987, in preparation). We simply note here that, except
for unambiguous cases of non-thermal emission, the emitting volumes may have sizes
$>$ to $>>$ 10^{15} cm. This is much larger than typical scales of "solar-type activity"
seen in X-rays or in the radio, pointing to a possible influence of the
circumstellar environment, which we now discuss.

4. CONTRIBUTION OF THE ENVIRONMENT TO NON-SOLAR ACTIVITY

4.1. Accretion disks: observational evidence

One of the most intriguing suggestions of contemporary stellar astronomy is
that stars may be surrounded by disks of dust and gas – which might be protoplane-
tary disks. Such disks are known to exist with certainty (imaging) in only a few
cases, like β Pic and similar stars (e.g. Sadakane and Nishida, 1986). It is
therefore conceivable that such disks already exist as a result of the star
formation process itself (see also below, § 4.2), i.e., are present around YSOs,
and detectable indirectly in a variety of ways.

In a recent survey of 12 T Tauri stars and related objects, Edwards et al.
(1987) did high-resolution spectra of forbidden emission lines of [SII] and [OI]
(sensitive to low-density regions) and of Hα emission. Modelling the line profiles, they found evidence for a mass outflow, likely in the form of a hollow cone.
In addition, since only the blueshifted part of the emission lines is seen,
Edwards et al. suggest, along the lines of Appenzeller et al. (1984), that
the receding part of the outflow is obscured by some optically thick material,
presumably a disk. In their view, therefore, there is accretion of mass from an
equatorial disk, which confines a bipolar mass flow in a hollow cone. We shall
come back to this model later (§ 4.3).

Using a different approach, Bertout (1987, and in preparation) has studied the
behaviour of an optically thick, warm disk around a T Tauri star. The disk is
assumed to be everywhere Keplerian, although the momentum loss by viscosity

results in a slow infall of material towards the center of the disk, i.e., in accretion towards the star. Following Lynden-Bell and Pringle (1974) (see also Pringle, 1981), the model is parametrized by the accretion rate, $\dot{M}_{acc}$, and features a hot boundary layer where the disk meets the accreting star. The temperature T_{BL} of this layer is $\approx$ 5 times the maximum disk temperature, $T_{d,max}$. For accretion rates $\approx$ a few 10^{-8} M_{Θ} yr^{-1} to a few 10^{-7} M_{Θ} yr^{-1}, Bertout (1987, in preparation) finds $T_{d,max}$ $\simeq$ 1500 to 2500 K, hence T_{BL} $\simeq$ 7500 to 12500 K. The total mass of the disk is $\approx$ 0.01-0.1 M_{Θ}. In this model, the disk therefore appears to emit UV and IR radiation in excess over the stellar spectrum: the UV comes from the boundary layer, the IR from distant regions of the disk. This gives indeed a good fit to the continuum spectra at several T Tauri stars (Fig. 8). Bertout (1987, in preparation) suggests, furthermore, than the prominent emission lines present (especially Hα) in the most active T Tauri stars come not from an active chromosphere, but from the boundary layer. This suggestion fits rather nicely the apparent absence of correlation between the X-rays and Hα emission (§ 3.1), and implies

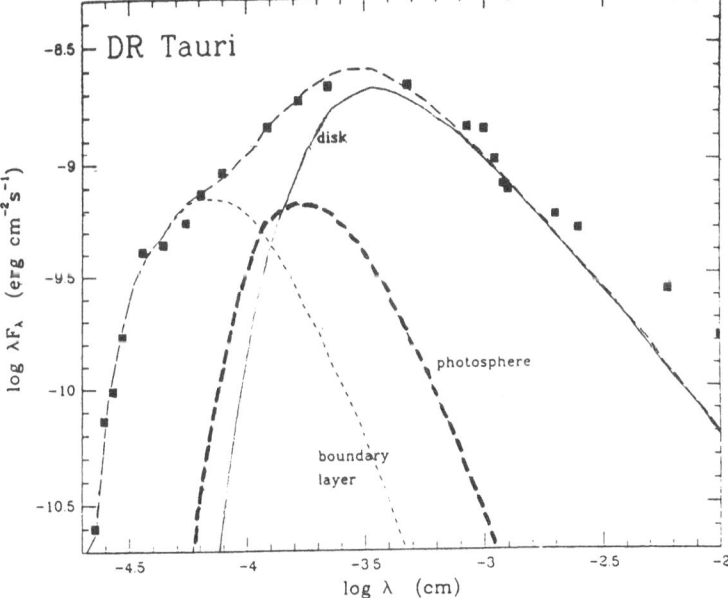

Fig. 8. The continuous spectrum of the active T Tauri star DR Tau, from the UV (IUE data) to the far-IR (IRAS data). In the model of Bertout (1987, in preparation), the spectrum is the superposition of the emissions of a warm disk (full line), of its boundary layer with the star (thin dotted line), and the photosphere itself (thick dotted line), which is seen in this case to be almost negligible!

that UV and IR excesses, in addition to strong optical and UV emission lines, should trace accretion disks. The odd case of DF Tau, in which there is evidence for a hot starspot at a high latitude (see § 2.2), could then correspond to magnetic accretion along a pole of the star, assumed to be an oblique rotator. We shall also come back to this model (§ 4.3).

4.2. <u>Towards a unified picture ?</u>

In view of the demonstrated evidence for outflow from YSOs, and of the possible evidence for accretion disks, one can build a unified time sequence of the early stages of stellar evolution (see, e.g., Adams et al. 1987). Briefly, the picture is as follows. A rotating clump of gas collapses in the form of a flattened disk. This disk further shrinks towards the central core, leading to a star surrounded by an accretion disk. Meanwhile, the interstellar magnetic field is tangled by the rotation at the disk, and either by a combination of magnetic twisting and centrifugal forces (see e.g., Pudritz and Norman 1986, Uchida and Shibata, 1987), or by an as yet unidentified stellar mechanism, a strong mass loss develops, in the form of jets and/or bipolar flows (Fig. 9).

Even if only a particular configuration of the magnetic field can give rise to and/or confine a bipolar flow, there may be accretion disks also with milder outflows (Edwards et al.) or without (Bertout). If the disk has a finite mass ≈ 0.01-0.1 $M_\odot$ as mentioned above, an accretion rate $\gtrsim 10^{-7}$ $M_\odot$ yr^{-1} gives a short timescale, $\approx 10^5$-10^6 yrs. The T Tauri star ceases to be active and becomes (?) a "post-" or "naked" (see Walter, 1986) T Tauri star.

4.3. <u>Objections and unsolved problems</u>

The above picture may have some truth, but the situation is still not satisfactory, for a number of reasons.

(i) The evolution described above cannot be a universal time sequence. For instance, the stellar radio sources in ρ Oph are certainly young (look at their spatial distribution in Fig. 3 of André et al., 1987), yet show no IR excess, hence, presumably, no warm disk.

(ii) On the contrary, classical active T Tauri stars are generally found at

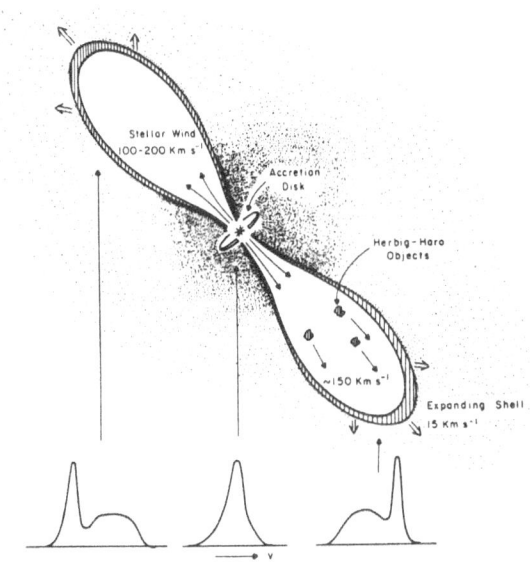

Fig. 9. An artist's view of the environment of a very young star, here applied to the case of L1551 in Orion (see Schwartz 1983). This consists in an accretion disk, an expanding, hollow shell observed in CO as a bipolar flow (see the line profiles at the bottom), and Herbig-Haro objects as tracers of the interaction between the flow and the molecular cloud. In some models, the flow is magnetized, and rotates along the rotation axis of the exciting star.

the edge of molecular clouds, and have ages > 10^6 yrs, i.e., greater than the presumed lifetime of a pristine accretion disk.

(iii) In the accretion disk + wind model of Edwards et al. (1987), the Hα emission is entirely due to the mass loss. The calculated $\dot{M}$ is ∼ 10^{-8} $M_\odot$ yr^{-1} in most cases, but rises to 10^{-6} $M_\odot$ yr^{-1} in the extreme cases of HL Tau and R Mon. However, no radio emission has been detected from these stars, down to very low levels (∼ 0.5 mJy), putting an upper limit < 10^{-8} $M_\odot$ yr^{-1} to their actual possible mass loss rate. Would Hα in that case be due to a boundary layer ?

(iv) From the radio data, cases of accretion are known around some PMS stars (T Tau (S), VS14 ?), but this implies hot disks (∼ 10^4 K), not warm disks.

It therefore seems that all the above examples of non-solar activity in YSOs do not yet fit into a single picture, even simplified. For instance, it is not even sure that a bipolar-flow phase should exist for all stars, as advocated by Lada (1985) ; only one such flow has been found in the ρ Oph cloud, for example (Walker et al. 1986, Fukui et al. 1986). Given the variety of objects considered, this may well be due to a difference in initial conditions: angular momentum, mass, ambient magnetic field, etc...

5. CONCLUSIONS

The few examples of present problems we have raised in the preceding section make this review more like a progress report. This basically reflects the fact that new data have been gathered at high speed, mainly in the new windows opened by space astronomy, and that to this day theory has just been unable to catch up.

One should however stress that, in spite of this wealth of data, a key ingredient in the study of stellar activity remains largely unknown: the magnetic field itself, of which only the strongest can be measured (see Borra et al., 1982), although major improvements have been made recently (Saar et al., 1986). In our opinion, this is one of the important reasons why the concept of the solar-stellar connection in general remains today largely qualitative.

The extension of this concept to young stellar objects in general (T Tauri stars remaining the most studied) has brought two main conclusions, however.

(i) X-ray, and, to a lesser extent, UV observations, have confirmed that YSOs display a conspicuous and widespread solar-type activity, mainly in the form of flares (note that weak underlying coronae cannot be excluded) ;

(ii) There is also a lot of evidence for a non-solar component of the activity, linked essentially to mass exchange with the surrounding medium: outflow (winds, jets, etc...), or inflow (accretion, either isotropic or from a disk).

The origin of the solar activity is most likely magnetic, as for other late-type stars (dynamo effect), although the correlations between activity tracers and stellar parameters are little more than trends. Again, the lack of knowledge of the magnetic fields is probably crucial. As for the non-solar activity, its origin is largely unknown. If it were confirmed that, at least in some well-defined cases, it is due only to environmental causes (e.g., accretion disks), then YSOs could be put more securely along with main sequence stars, to widen the available range of stellar parameters. For instance, it would be very interesting to systematically compare a set of cool, weakly active PMS stars, still on their convective track, all at the same spectral type (hence at different ages, since these tracks are almost vertical), with main sequence stars of exactly the same spectral type, e.g., $\approx$ MO or later. For this type, luminosities go from ≈ 3 $L_\odot$ down to ≈ 0.03 $L_\odot$, the v sini span the range ≈ 25 km.s^{-1} to less than a few km.s^{-1}, etc... Other ongoing studies are also of course worth pursuing.

Beyond the problem of the validity of the "solar-stellar connection", we have at least learned that intense activity (solar as well as non-solar) is in fact a characteristic of very young stars. Therefore, activity studies, in the broadest sense, remain a prime concept to trace the pre-main sequence phases of stellar evolution.

ACKNOWLEDGMENTS

It is a pleasure to thank M. Schüssler and the Organizing Committee in Titisee for their kind hospitality, and Philippe André for useful discussions.

REFERENCES

Adams F.C., Lada C.J., Shu F.H., 1987, Ap. J. 312, 788.
André Ph., 1987, in Montmerle and Bertout (1987).
André Ph., Montmerle T., Feigelson E.D., 1987, Astr. J. 93, 1182.
Appenzeller I., Jankovics I., Ostreicher R., 1984, Astr. Ap. 141, 108.
Basri G., 1987a, Ap. J., 316, 377.
Basri G., 1987b, in Bertout and Montmerle (1987).
Bertout C., 1983, Astr. Ap., 126, L1.
Bertout C., 1984, Rep. Prog. Phys. 47, 111.
Bertout C., 1987, in Circumstellar Matter, eds. I. Appenzeller and C. Jordan
   (Dordrecht: Reidel), in press.
Borra E.F., Landstreet J.D., Mestel L., 1982, Ann. Rev. Astr. Ap. 20, 191.
Bouvier J., 1987a, Ph. D. Thesis, Université Paris VII.
Bouvier J., 1987b, in Montmerle and Bertout (1987).
Bouvier J., Bertout C., Benz W., Mayor M., 1986, Astr. Ap. 165, 110.
Cabrit S., Bertout C., 1986, Ap. J. 307, 313.
Cabrit S., Bertout C., 1987, in Montmerle and Bertout (1987).
Casoli F., 1987, in Montmerle and Bertout (1987).
Catala C., Felenbok P., Czarny J., Talavera A., Boesgaard A.M., 1986, Ap. J.,
   308, 791.
Cohen M., Bieging J.H., 1986, Astr. J. 92, 1396.
De Campli W.M., 1981, Ap. J. 244, 124.
Dulk G.A., 1985, Ann. Rev. Astr. Ap. 23, 169.
Edwards S., Cabrit S., Strom S.E., Heyer I., Strom K.M., Anderson E., 1987, Ap.
   J., in press.
Feigelson E.D. 1984, In Cool Stars, Stellar Systems, and the Sun, ed. S. Baliunas
   and L. Hartmann (Berlin: Springer).
Feigelson E.D., 1987, in Montmerle and Bertout (1987).
Feigelson E.D., De Campli W.M., 1981, Ap. J. (Letters) 243, L89.
Feigelson E.D., Montmerle T., 1985, Ap. J. (Letters) 289, L19.

Fukui Y., Sugitani K., Takaba H., Iwata A., Mizuno H., Ogawa H., Kawabata K. 1986, Ap. J. (Letters) 311, L85.

Gilman P.A., 1983, in Solar and Stellar Magnetic Fields, Origin and Coronal effects, ed. J.O. Stenflo (Dordrecht: Reidel), p. 247.

Guilloteau S., 1987, in Montmerle and Bertout (1987).

Hartmann L., Edwards S., Avrett E., 1982, Ap. J. 261, 279.

Hartmann L.W., Soderblom D.R., Stauffer J.R., 1987, Astr. J. 93, 907.

Hjellming R., Gibson D. (Eds.), 1985, Radio Stars (Dordrecht: Reidel).

Klein K.L., Chuideri-Drago F., 1987, Astr. Ap. 175, 179.

Klein K.L., Trottet G., 1984, Astr. Ap., 141, 67.

Lada C.J., 1985, Ann. Rev. Astr. Ap. 23, 267.

Lago M.T.V.T., 1982, M.N.R.A.S. 198, 445.

Lynden-Bell D., Pringle J.E., 1974, M.N.R.A.S. 168, 603.

Montmerle T., Bertout C. (Eds.) 1987, Protostars and Molecular Clouds (Saclay: CEA/Doc), in press.

Montmerle T., Koch-Miramond L., Falgarone E., Grindlay J., 1983, Ap. J. 269, 182.

Montmerle T., Koch-Miramond L., Falgarone E., Grindlay J., 1984, Phys. Scripta T7, 59.

Mundt R., 1981, Astr. Ap. 95, 234.

Mundt R., Bührke T., Fried J.W., Neckel T., Sarcander M., Stocke J., 1984, Astr. Ap. 140, 17.

Pallavicini R., 1985, in Hjellming and Gibson (1985), p. 197.

Panagia N., Felli M., 1975, Astr. Ap. 39, 1.

Pringle J.E., 1981, Ann. Rev. Astr. Ap. 19, 137.

Pudritz R.E., Norman C.A., 1986, Ap. J. 301, 571.

Reynolds S.P., 1986, Ap. J. 304, 713.

Rosner R., Golub L., Vaiana G.S., 1985, Ann. Rev. Astr. Ap. 23, 413.

Saar S.H., Linsky J.L., Beckers J.M., 1986, Ap. J. 302, 777.

Sadakane K. Nishida M., 1986, Pub. Astr. Soc. Pac. 98, 685.

Schwartz R. D., 1983, Ann. Rev. Astr. Ap. 21, 209.

Ushida Y., Shibata K., 1987, in Star Forming Regions, eds. M. Peimbert and J. Jungaku (Dordrecht: Reidel), p. 385.

Walker C.K., Lada C.J., Young E.T., Maloney P.R., Wilking B.A., 1986, Ap. J. (Letters) 309, L47.

Walter F.M., 1986, Ap. J. 306, 573.

Walter F.M., Kuhi L.V., 1981, Ap. J. 250, 254.

Walter F.M., Kuhi L.V., 1984, Ap. J. 284, 194.

WINDS IN LATE TYPE STARS AND THE SOLAR WIND

D. Reimers
Hamburger Sternwarte, Universität Hamburg
Gojenbergsweg 112, D-2o5o Hamburg 8o, F.R.G.

I. Introduction

Both the solar wind and winds in late type giant stars have been studied since more than 25 years. While the solar wind was postulated by Ludwig Biermann in 195o from the fact that comet tails always point away from the Sun and was probed for the first time by Mariner 2 in 1962, mass-loss of a late type giant was discovered when Deutsch (1956) found blue shifted resonance lines in the spectrum of the close visual companion of the M5II giant α Her.

In the first two decades the focal points of research on the solar wind and on cool giant winds were quite different. For cool stars, the main motivation for wind studies was to determine the rates of mass-loss with the aim to understand late stages of stellar evolution. The solar wind, on the other hand, has been studied with in situ measurements of the solar wind plasma in the ecliptical plane at various distances from the Sun ($\sim$ o.3 A.U. to $\gtrsim$ 15 A.U.). This has led to a detailed picture of the dynamics and of the plasma properties of the solar wind. Only recently - with the advent of X-ray and UV satellites - physical conditions of chromospheres, coronae, and winds of cool giants could be studied, and only on a large scale with low geometrical resolution. The detection of stars like the "hybrid giants" with properties of the outer layers common to the Sun (hot corona and fast wind) and to cool supergiants (heavy mass-loss) has been a challenge to both researchers working on the solar wind and on winds of cool giants.

While solar wind and coronal studies may help to identify complex physical processes important also in stars, late type giants with properties not too different from the Sun (e.g. G and K giants and supergiants) may help to identify mechanisms not yet understood in the Sun like the heating of the solar corona. What would happen to the solar wind if we could "blow up" the Sun to $\sim$ 4o solar radii. Do observations of the wind of a G supergiant give the answer?

One thing this review might accomplish is to demonstrate on one hand the uncomplete picture of stellar winds we have - all information is from a few spectral lines - and to point out the tentative nature of our present understanding of the outer layers of the Sun. We should not be too optimistic about theoretical stellar wind models as long as the solar wind is so poorly understood in spite of detailed in situ measurements.

In section II, I summarize briefly the present state of knowledge of stellar winds for those stars most relevant for a comparison with the Sun, i.e. G, K and M giants and supergiants. Winds of the coolest luminous stars like OH-IR stars, in which we observe dust, maser emission and thermal emission from molecules, will not be discussed here since in these far evolved stars most probably other mechanisms like radiation pressure on dust particles are effective.

In section III, I try to summarize very briefly those - large scale - aspects of solar wind observations which appear relevant for a comparison with winds in late type giants.

A comparison of mass-loss rates, wind velocities, energies required to drive the winds, and energy losses of the outer layers of cool stars, including the Sun, is made in section IV.

II. Winds of cool stars

1) Mass-loss indicators and incidence of mass-loss in the HR diagram

a) Circumstellar CaII, MgII and Ly α: All stars cooler and more luminous than on a line in the HR diagram defined roughly by (K5, M_V = o), (K4, -1), (K2, -1.8), (G5, -4), (Go, -4.5) have far-shifted CaII H+K absorption components (Reimers, 1977a), cf. Fig.1. In case of the spectroscopic binaries µUMa (MoIII) and ξ Cyg (K5Ib) it could be shown that these K4 absorption lines remain stationary - while the photospheric lines move back and forth due to orbital motion - and thus indicate circumstellar matter.

Similar far-shifted absorption lines have been found in the MgII resonance lines (Hartmann et al., 1981; Reimers, 1982, Dupree and Reimers, 1987). Due to larger optical depths of the MgII lines (higher Mg abundance) compared to CaII lines the MgII visibility limit is at slightly lower luminosity and higher temperatures. Line asymmetries most probably caused by winds have also be seen in Ly α, in α Aur, and α TrA (cf. Dupree and Reimers, 1987). In the CaII and MgII resonance doublets, the ratio of the violet to the red emission peak (V/R) provides more indirect evidence for outflow of matter from red giants. Stencel (1978) and Stencel and Mullan (198o) have shown that the lines bounding those regions in the HR diagram where V/R > 1 (no outflow) changes to V/R < 1 (outflow) are near to and parallel to the line in the HR diagram beyond which stars have CS lines. It is not clear whether V/R < 1 values are just caused by blue-shifted absorption by escaping material (far away from the star) or are caused by expanding chromospheres, or both. Without detailed modelling, the origin of the asymmetry of the selfreversed CaII and MgII profile remains unclear. Even the conclusion that V/R < 1 means mass-loss appears still speculative.

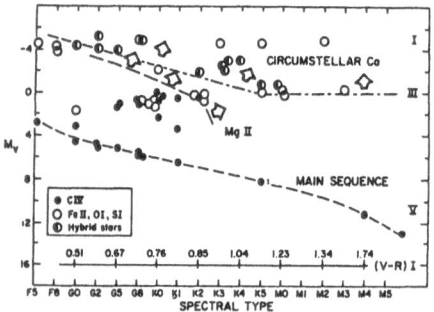

Figure 1. The presence of various spectral lines in stars of different spectral types and luminosity classes. The broken line denotes the boundary above which variable CS CaII lines appear (Reimers, 1977). The position of the MgII asymmetry change is also indicated (Stencel and Mullan, 198o), from Dupree and Reimers (1987).

b) HeI 1o83o Å: Among K giants there is evidence for a nearly one to one correlation between the presence of HeI emission and a CaII K asymmetry ratio V/R < 1 (see below). A number of K giants and G supergiants (ι Aur, α Aqr, θ Her, γ Aql) have far-blue-shifted (-15o to -2oo km/s) HeI 1o83o Å absorption components (O'Brien and Lambert, 1986).

c) Hα-emission: It is known since the work of Kraft et al. (1964) and Gahm and Hultquist (1972) that in G and K supergiants there is variable Hα-emission - mostly blue-shifted - superimposed upon the broad chromospheric Hα absorption line. Similar Hα-emission components that have been considered as due to mass loss have been found in bright Population II stars (e.g. Cohen, 1976; Cacciari and Freeman, 1981), and in extremely luminous F and G supergiants. Severe doubts have been cast on former simple interpretations of the Hα-emission in terms of mass-loss in a recent paper by Dupree, Hartmann and Avrett (1984); and much lower mass-loss rates were found.

d) In K supergiants like λ Vel, the emission cores of CaII H and K are blue-shifted. The stellar winds in red giant stars seem to be under way already in the chromospheres with typically half the terminal velocity (Wilson, 196o; Reimers, 1975a, Fig.2). In α Ori, the Hα absorption core - formed in the extended chromosphere of the star - is also blue-shifted by about half the wind terminal velocity. In addition, the absorption core does not follow the photospheric pulsation, indicating that the chromosphere is well decoupled from the photosphere (Goldberg, 1979).

e) Radio emission: A deep survey of the nearest, cool giants and supergiants has been conducted with the VLA at 6 cm (Drake and Linsky, 1986). Only four K and M giants (α^1 Her, α Boo, ρ Per, μ Gem) have been detected definitely. Radio emission of giant stars at 6 cm is thermal emission from cool, partly ionized winds. With estimates of the degree of ionization ($n_e/n_H \approx$ 0.1 ... 0.01) and spectroscopically determined wind velocities the mass-loss rates can be estimated.

In a binary system consisting of an M or K supergiant and a hot B star, the latter may form an HII region within the wind of the supergiant which is observable at cm-wavelength. Systems studied quantitatively are the GoIa supergiant HR8752 (Smolinski et al. 1977; Lambert and Luck, 1978) and α Sco (Hjellming and Newell, 1983). Further detected systems are 31 Cyg and 47 Cyg (Drake, Brown and Reimers, in preparation). Very recently, the CS envelope of α Ori could be detected in the HI 21 cm line (Bowers and Knapp, 1986).

f) Emission from dust and molecules: All sufficiently luminous and cool supergiants, Mira stars etc. with high mass-loss rates have thermal emission from dust and from abundant molecules like CO as well as maser line emission of OH, H_2O, etc.

This review will mainly concentrate on a comparison of winds of less luminous and somewhat hotter red giants with the solar wind. Therefore, the advanced evolutionary stages with the highest mass-loss rates will not be discussed here.

g) Cool stars with hot companions: The wind of a red giant can be seen in strong resonance lines in the spectrum of a close visual companion e.g. in α^2 Her, α Sco B, and in Mira B.

With the launch of IUE, the optical separation of red giants with hot companions could be replaced by a separation through complementary energy distributions of the components. As will be shown below, this technique has yielded the most detailed wind studies and among the most accurate mass-loss rates available for red giants.

2) Wind velocities

In red giants, wind velocities are usually below the velocity of escape from the stellar surface. A survey of stellar wind velocities in the whole red giant region of the HR diagram has been made by means of CaII K_4 shifts (Reimers, 1977 a), while a compilation of wind velocities from MgII lines is given by Dupree and Reimers (1987).

A trend can be seen from luminous M supergiants ($\approx$ 10 km/s) over early M giants (25 km/s) and G and K (Super)-giants ($\approx$ 50 to 150 km/s) to the solar wind ($\sim$500 km/s). Detailed measurements of wind velocities by means of high-resolution IUE spectra taken at various binary phases of ζ Aur, 31 Cyg, 32 Cyg, 22 Vul and δ Sge confirm the values found more simply from K_4 shifts.

Well studied stars with good estimates of both wind velocity v_w and velocity of escape from the stellar surface are given in Table 2 and Fig. 2. Results are discussed in section IV 1.

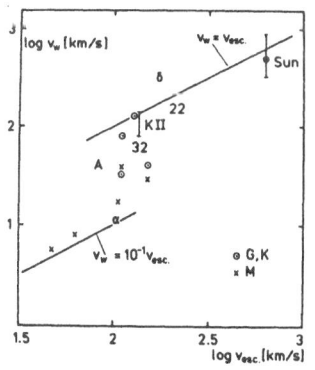

Figure 2. Stars with well determined wind velocities v_w (= asymtotic flow speeds) and gravitational escape velocities v_{esc} from Table 2. α = α Ori, A = α Boo, δ = δ And, 22 = Vul, 32 = 32 Cyg.
The upper and lower lines correspond to $\dot{E}_{kin} = \dot{E}_{pot}$ and $\dot{E}_{kin} = 10^{-2} \dot{E}_{pot}$ for the winds.

3) Dividing lines, "hybrid atmosphere" stars, wind variability

A mentioned above, circumstellar CaII H and K and MgII h and k lines are seen to the right of a line in the HR diagram which runs from about MoIII through the K giants to early G supergiants (c.f. Fig. 1). After the first IUE observations, it was suggested

(Linsky and Haisch, 1979) that among red giants there is a sharp division in the HR diagram between "solar like" stars that emit transition layer lines (NV, CIV, SiIV, ...) and "non solar like" stars, that emit only low temperature, chromospheric lines (OI, SI, CI, SiII, ...), and that this seemingly sharp line coincides with the lower boundary of stars in the HR diagram with observed cool stellar winds (Reimers, 1977a). X-ray observations of cool giants with the Einstein-Satellite (HEAO-B) seem to confirm the existence of a dividing line between stars with and without detectable X-ray emission (e.g. Linsky, 1981), and occurring at a similar location in the HR diagram. The most obvious interpretation was that the onset of massive winds inhibits the formation of a hot "corona" through efficient cooling.

However, it has become clear later, that the transition in the HR diagram from stars with "solar type" outer atmospheres (transition layer line and X-ray emission) to stars with cool, massive winds is not sharp and that, in particular, "mixed" types occur, the so-called "hybrid atmosphere" stars.

These stars are early G supergiants like α Aqr, β Aqr, δ TrA or K giants of luminosity class II (e.g. θ Her, γ Aql, ι Aur, α TrA), all close to the mentioned wind/corona boundary line (c.f. Fig.1). Hybrid stars have both a cool, high-velocity wind (55-180 km/s) and transition layer line emission (CIV, CIII, CII, NV).

In Table 1 I have collected data for all 12 presently known hybrid stars. Spectra of two new hybrid stars (not yet published) μ UMa and σ Oph, are shown in Fig.3.

σ Oph (HD 157999) has highly variable CS CaII H_4+K_4 components which disappear sometimes. σ Oph is a SB for which the CS CaII H+K lines could be shown to remain stationary at -75 km/s while the photospheric lines move back and forth (Reimers, 1977a). The star has a far shifted HeI 10830 Å line (up to -165 km/s) like the hybrid stars γ Aql, ι Aur, θ Her and α Aqr (O'Brien, 1980), The UV spectrum does not reveal the companion, which accordingly must be a late main sequence star or a fairly cool white dwarf. According to the high wind velocity, the transition layer lines, and HeI 10830 Å, μ UMa is a typical hybrid star in spite of its late spectral type (MoIII).

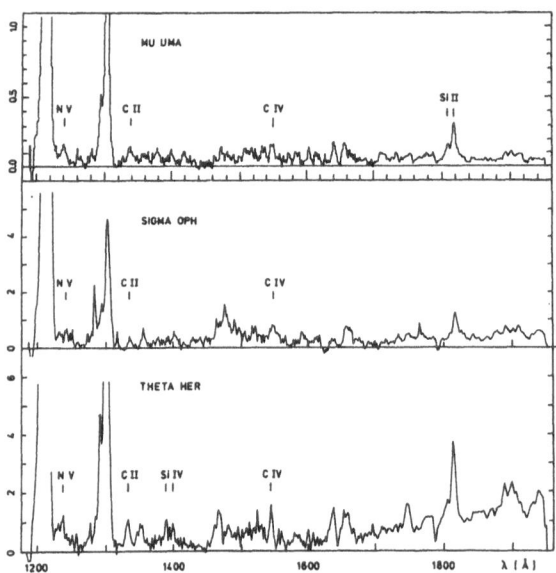

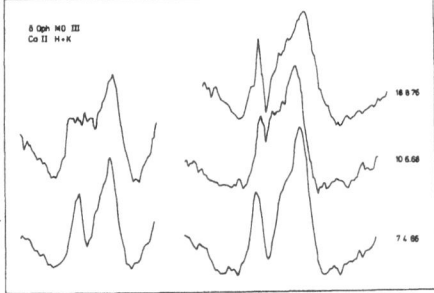

Figure 4. Variable circumstellar CaII H (left) and K (right) lines in δ Oph (MoIII).

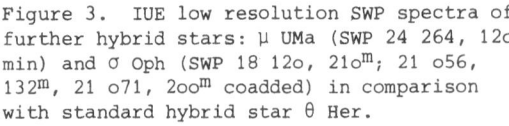

Figure 3. IUE low resolution SWP spectra of further hybrid stars: μ UMa (SWP 24 264, 120 min) and σ Oph (SWP 18 120, 210^m; 21 056, 132^m, 21 071, 200^m coadded) in comparison with standard hybrid star θ Her.

Four of the hybrid stars have been observed for X-ray emission with the Einstein X-ray observatory. For α Aqr, β Aqr and γ Aql, only upper limits of $f_x/f < 0.3$-$1 \cdot 10^{-7}$ were found (Haisch and Simon, 1982), while θ Her is a marginal detection with

$f_x/f = 1.4 \cdot 1o^{-7}$ which corresponds to $\log L_x \sim 29.8$ (Drake et al. 1984a).

EXOSAT observed 3 hybrid stars and detected α TrA (Brown, 1986). According to Brown, the detection of α TrA suggests the presence of plasma at Te $> 1o^6$ K with emission from cooler regions being a less likely explanation.

According to the common occurrence of hybrid stars near the corona/wind dividing line, the concept of a sharp transition from one modus (hot corona, no detectable wind) to another (massive, cool wind, extended chrom., no hot plasma) is no longer tenable.

One characteristic of the hybrid stars and of other stars close to the dividing line is the variability of CS CaII lines (Reimers, 1977). For several hybrid stars, variability of circumstellar MgII lines has been studied and detected in α Tra, β Aqr, γ Aql (c.f. Dupree and Reimers, 1987).

In a study of variability of UV emission lines (mainly chromospheric) of red giants and supergiants, Oznovich and Gibson (1986) found variability in 4 stars ($\triangleq$ 15% of their stellar set): α Aqr, β Peg, σ Oph, and γ Aql. Three of these stars are hybrid stars.

HeI 1o83o $\overset{o}{A}$ emission is also variable. In Θ Her, e.g., HeI emission showed cyclic variations with maximum in 1966, 1971 and 1975 (Zirin, 1976). The characteristic time scale for larger variations of CS lines in K giants is of the order of one to several months. Examples are α Tau, μ UMa, γ Aql, θ Her, and 63 Cyg (Reimers, 1977a, 1982), cf. Fig.4.

Hybrid stars, in particular the K bright giants of L.C.II, have been shown to have usually an excess CaII H+K flux, indicative of a high activity level (Middelkoop, 1982). From the occurrence of KII giants in clusters of intermediate turnoff mass ($\geq$ 4 M$_\odot$) it appears plausible that these stars have quite recently evolved off the main sequence and rotate still faster than normal (L.C.III) giants. In case of 22 Vul (G3II-Ib), which is in many respects similar to hybrid stars (Reimers and Che-Bohnenstengel, 1986), a rotational velocity of 18 km/s has been measured.

Table 1. Known Hybrid Stars

Star	HD	Sp.	M_v	v(CaII) [km/s]	v(MgII)		Ref.	Comment
β Aqr	2o4867	GoIb	-4.7	-8o	$-$ 75....-135		(1)	
α Aqr	2o975o	G2Ib	-5.3		-127		(1)	
δ TrA	145544	G2II	-3.9	-88	$-$ 81		(3)	
Θ Her	16377o	K1IIa	-2.5	-69	$-$ 75....-1oo		(3)	X-rays?
α TrA	15o798	K2IIb -IIIa	-2.4	94	$-$128	$-$18o	(2)	X-rays,binary?
γ Aql	186791	K3II	-1.9	-76 var.	$-$ 7o....-11o		(4)	
ι Aur	31398	K3II	-1.5	-75	$-$ 77		(3)	
σ Oph	157999	K2II	-1.8	-95 var.			this paper	
	81817	K3III	-2.3	variable		$\}$	(5)	WD companion
q Car	-	K3IIa					(6)	
δ And	3627	K3III			-3oo		(7)	Binary
μ UMa	89758	MoIII	-1	-55 var.			this paper	Binary

References:

(1) Hartmann et al. (1981a)
(2) Hartmann et al. (1981b)
(3) Reimers (1982)
(4) Hartmann et al. (1985)

(5) Reimers (1984)
(6) Brown (1986)
(7) Judge et al. (1986)
For further candidates cf. Brown (1986)

If winds of hybrid stars represent something like the missing link between hot, solar type winds and cool M giant winds, the question arises, whether there is evidence for a warm wind of $\sim 10^5$ K in these stars.

Hartmann et al. (1981) proposed that the broad transition region emission lines (e.g. CIV, CIII and SiIII lines) are formed in the wind, whereas an alternative would be that the wind originates in magnetically open regions, while the hot plasma is confined to closed magnetic regions.

High dispersion, wavelength calibrated IUE spectra of α Tra and γ Aql show that the SiIII 1892 Å and CIII 1908 Å intersystem lines are essentially at their rest wavelengths ($|\Delta v| \leq 5$ km/s) (Brown et al. 1986). This means that these transition region emission lines, which are formed at temperatures near $5 \cdot 10^4$ K, are formed in stationary or only very slowly expanding regions. There is no evidence for hot gas expanding at velocities comparable to the wind velocities observed in MgII or CaII (cf. Table 1).

The widths of the CIII and SiIII lines in αTrA implies nonthermal broadening (turbulence) with a velocity ~ 30 km/s (Brown et al., 1986).

In the G2II-Ib star β Dra, which however, has no detectable wind, there is even evidence for downflow in SiIII 1982 relative to low excitation chromospheric lines (Ayres et al., 1986).

3. Mass-loss rates

a) Optical CS lines of single stars

While the winds, or circumstellar shells, of late type giants are easily detected as violet shifted cores - at high resolution P Cyg profiles superimposed upon photospheric line cores - of strong resonance lines of neutral or singly ionized metals like CaII, MgII, TiI and TiII, BaII, SnII, NaI, FeI, ..., it has turned out to be impossible to determine mass-loss rates quantitatively from these lines. The reason is that while it is possible to measure ion column densities N_{ion} and wind velocities V_w from a theoretical analysis of the P Cyg type lines, it is not possible to infer from spectroscopic observations where in the line of sight the CS lines are formed. Since the mass-loss rate $\dot{M} \alpha N_{ion} \cdot v_w \cdot R_i$, where R_i is the inner shell radius, and $R_i \neq R_{star}$ is not known, $\dot{M}$ cannot be determined from CS lines of single M giants and supergiants.

The only technique for measuring mass-loss of single (nonbinary) stars appears to be spatially resolved imaging of CS shells in scattered resonance line photons like KI 7699 Å or NaD (Mauron et al., 1984; 1986). In case of α Ori, the KI line was seen as far as 50" from the star.

However, since KI and NaI are minor ionization species, reliable knowledge of the ionization of metals and of the formation of CO (carbon can be a major electron donator) is necessary. Nonequilibrium effects (flow time $\sim$ recombination time scale in the outer envelope) could be shown to be negligible for KI while large effects have been found for α Ori for CaI, CaII and MgI at distances $\geq 10^3$ stellar radii (~ 20" apparent distance) from detailed ionization calculations including outflow effects (Robel, 1987).

b) Microwave continuum emission

The partly ionized winds of K and M giants reveal thermal ff-emission. Four nearby giants have been positively detected at 6 cm and 2 cm with the VLA with spectral indices close to the 0.6 ($S_v \sim v^{0.6}$) as predicted for an optically thick wind (Drake and Linsky, 1986). The rate of loss of ionized matter for α Her is ~ 1 % of the total rate as determined by the binary technique (Reimers, 1977b). If similar ionization degrees are valid for K giants like α Boo, a mass-loss rate of $\sim 7 \cdot 10^{-9}$ M$_\odot$/yr is derived. This is certainly an upper limit, since ionization degrees could be higher than in α Her (M5II). From chromospheric modelling using MgII emission, Linsky (1986) gives $\dot{M} = 2 \cdot 10^{-10}$ M$_\odot$/yr and an ionization fraction of ~ 50 % for α Boo.

c) Binary technique

The rate of mass-loss can be determined from CS absorption lines of a predominant stage of ionization seen in the spectrum of either a visual companion or of a hot

companion which can be separated from the red giant in the UV due to its complementary spectral energy distribution. In both cases one avoids the difficulty of locating
the shell, since the geometry of the visual system or the binary system with known orbital elements permits to locate the origin of the CS lines.

The visual binary technique has been applied to α Her (Deutsch, 1956; Reimers,
1977b), α Sco (Kudritzki and Reimers, 1978) and to O Cet (Reimers and Cassatella,
1985b).

With the launch of the IUE, the UV technique could be applied to a number of ζ Aur
systems (eclipsing binaries) and VV Cep systems (like α Sco, Boss 1985). At IUE wavelengths, the optical separation of red giants with hot companions can be replaced by
a separation through complementary energy distribution of the components. At IUE wavelengths, in particular in the short-wavelength range, one observes a pure B star spectrum upon which numerous CS P Cygni type lines formed in the extended wind and chromospheric absorption lines (near eclipse) of the red giant are superimposed.

The B star serves as an astrophysical light source (a "natural satellite") which
moves around in the wind of the red giant. However, compared to widely separated
visual binaries, a number of additional difficulties arise
- a non-spherical, 3-dimensional line transfer problem has to be solved since the
light source (B star) is excentric from the wind symmetry center. Computer codes that
solve this problem have been developed in the 2-level approximation by Hempe (1982,
1984) and for the multilevel case by Baade (1986)
- the wind is disturbed in the immediate surrounding of the B star as it moves super-
sonically through the wind and formes an accretion shock front (Chapman, 1981). However, a detailed study of the accretion shocks has shown that their geometrical size
is very small compared to the CS shell and can be neglected in line transfer calculations (Che-Bohnenstengel and Reimers, 1986)
- the hot B star ionized the wind, i.e. an HII region is formed within the red giant
wind. In 31 Cyg and α Sco, in particular, the size of the HII region is large, and
it has to be taken into account quantitatively since ions like SiII and FeII which
are used for the mass-loss rate determination may be doubly ionized within the HII
regions.

On the other hand, ζ Aur binaries are the only stars besides the Sun where the
winds and extended chromospheres can be studied with spatial (height) resolution.

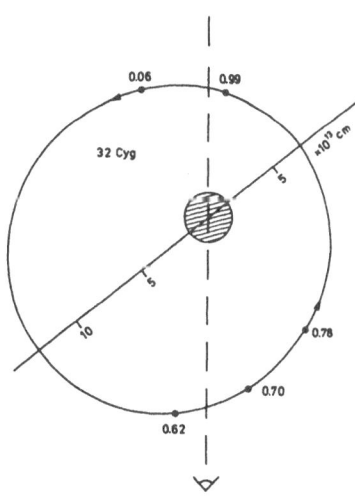

Figure 5 a) Location of B star
relative to K supergiant at observed phases.

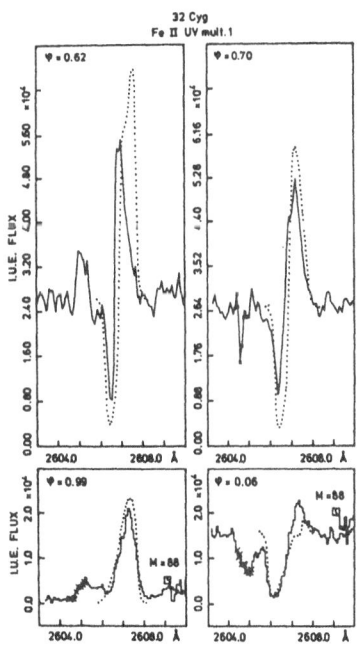

Figure 5 b) Dependence of a wind
line (FeII UV Mult.1) on phase in
comparison with theory.

In ζ Aur/VV Cep systems, the wind is visible at all phases in P Cyg type profiles (during total eclipse of B star pure emission lines) of ions like FeII, SiII, SII, MgII, CII, AlII, and oI. These lines are formed by scattering of B star photons in the wind of the red giant. A few wind lines like FeII Mult.9 (∿ 1275 Å) are seen in pure absorption due to the branching ratios of the upper levels which favour re-emission as FeII UV Mult. 191 photons (Hempe and Reimers, 1982; Baade, 1986).

Theoretical modelling of wind line profiles and of their phase dependency has yielded accurate mass-loss rates and wind velocities for a number of systems (Table 2). It has turned out that a good mass-loss determination requires both phases with the B star in front of the red supergiant (which yields wind turbulence v_t) and phases with the B star behind the red supergiant (which yield the wind velocity v_w). Typically, $v_w \approx 2 v_t$. Further details can be found in Che et al. (1983). It turned out that it was possible to match the circumstellar line profiles at all phases with one set of parameters v_w, v_t and - within a factor of 2 - one mass-loss rate $\dot{M}$ (Figs.5,6). This means that at least in the orbital plane the envelope asymmetries (in density) are within a factor of 2 on a length-scale of several K giant radii.

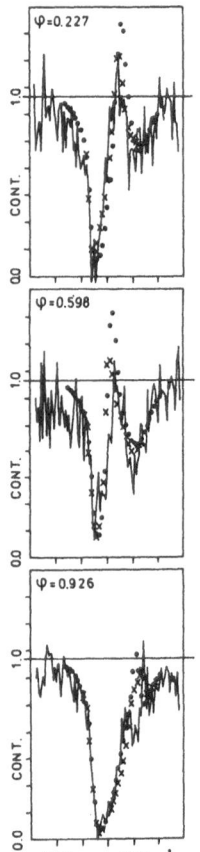

Figure 6. Observed and theoretical SiII UV Mult.1 line at different phases for 22 Vul (G3Ib-II), from Reimers and Che-Bohnenstengel.

In several binary systems consisting of a red giant and a hotter companion, the B star is early enough (earlier than ∿ B 4) to ionize the wind in part. The HII region within the wind produces thermal Bremsstrahlung which has been detected in a number of systems (cf. section II1). Mass-loss rates have been determined for α Sco (Hjellming and Newell, 1983) and HR 8752 (Lambert and Luck, 1978). A compilation of the most accurate mass-loss rates of red giants is given in Table 2 and Fig.11.

4. Extended chromospheres and wind expansion

There is growing evidence that closely related to strong mass-loss in red giants is the existence of geometrically extended chromospheres.

During a lunar occultation of 119 Tau (M2Iab) it was observed that H_α light comes from a region having at least twice the diameter of that which produces the continuum. In α Ori it was observed by means of speckle interferometry in H_α that the detectable H_α diameter is about 5 times the stellar diameter (Goldberg et al., 1981; Hege et al., 1986). It was also found that the extended H chromosphere has an elongated shape. Recall that during a solar eclipse bright red features (H_α) are seen here and there around the obscured disk. This was the origin of the name "chromosphere" of the Sun.

In Zeta Aurigae type eclipsing binary systems, the extended chromospheres can be studied with height resolution when the B star moves behind the extended atmosphere of the K supergiant. IUE observations with high spectral resolution during chromospheric eclipse phases offer several advantages over optical studies as performed in the 1950s (cf. Wilson, 1960b): (i) in the UV at wavelengths λ ∿ 2800 Å, the B star provides a smooth continuum on which the chromospheric absorption lines are superimposed (ii) in strong UV lines the chromosphere can be seen up to projected heights of about one stellar radius (iii) the steep increase of continuous opacity - Rayleigh scattering at neutral hydrogen - provides a means to derive a density model of the inner chromosphere from the wavelength and time dependence of totality in UV below 2ooo Å.

Schröder (1985a, 1985b, 1986) has used both absorption lines and continuum data for constructing empirical model chromospheres for 32 Cyg, ζ Aur and 31 Cyg. Chromospheric densities could be represented by power laws of the form $\rho \sim r^{-2} \cdot h^{-a}$ with $a \approx 2.5$ where r is the distance from the center of the star and h is the height above the photosphere. The empirical density distribution shows that after a steep decrease in the inner chromosphere already in the upper chromosphere (height > 1/2 to 1 giant star radii $R_{::}$) expansion starts ($\rho \sim r^{-2}$), a typical density at a height of 2 $R_{::}$ is 10^7 cm^{-3} (Fig.7).

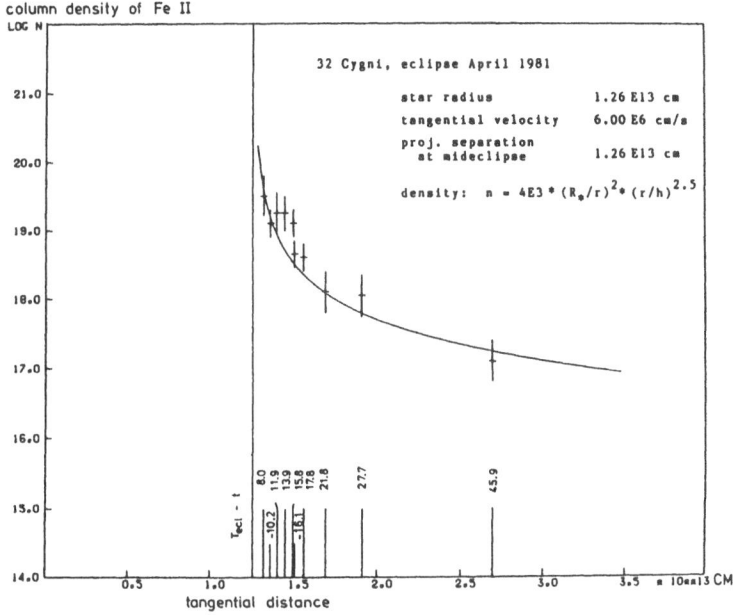

Figure 7. Chromospheric density distribution obtained from curve of growth analysis of ultraviolet FeII lines (Schröder, 1985a,b).

Since one observes total particle densities in the expanding chromosphere up to $h \approx 1.5\ R_K$, and in addition the wind density and velocity outside of 5 R_K, one can try to look for consistency by assuming a steady wind, i.e. to apply the equation of continuity. Using $\dot{M} = 4\ \pi\ r^2 \cdot \rho(r) \cdot v(r)$ and $\rho(r) = \rho_0 \cdot (R_{::}/r)^2 \cdot (r/(r-R_{::}))^a$ for the chromosphere, we find $v(r) = \dot{M}\ (4\ \pi\ \rho_0\ R_{::}^2)^{-1}\ (1 - R_{::}/r)^a$ and a wind terminal velocity $v_\infty = \dot{M}\ (4\ \pi\ \rho_0 \cdot R_{::}^2)^{-1}$ which can be checked with observed values for ρ_0, $\dot{M}$ and v_∞ for consistency. For 32 Cyg and 31 Cyg Schröder (1985) found consistency, which means that the empirical density distribution (when extrapolated by the equation of continuity to the outer wind) yields the correct mass-loss rate. In case of ζ Aur, the chromospheric density distribution was far too steep - at least at that particular limb position during eclipse - to give the mean mass-loss rate, which might be a stellar analogue to a solar coronal hole.

Another stellar-solar analogue is the prominence detected during egress of the 1981 eclipse of 32 Cyg (Schröder, 1983). After egress from eclipse, an additional 'dip' in the light curve was seen at wavelengths λ < 2000 Å for at least 6 days. Since the observed "prominence" was optically thin, the observed frequency of optical depths $\tau_\nu \sim \nu^{5.5}$ could be used to identify the opacity as Rayleigh scattering at HI ground state. A linear extension - perpendicular to the line of sight - of about 1/6 K giant radii ($\approx$ 30 $R_\odot$) and an apparent height of $\sim$ 15 $R_\odot$ above the limb (at 1350 Å) was estimated from the light curves. The small observed velocity of +20 km/s as measured from a few absorption lines like VII 3110.7 or TiII 3072 Å seen in addition to the normal chromospheric lines indicates a slowly moving cloud. Also, even a moderate velocity perpendicular to the line of sight, e.g. a slow prominence moving upwards with the wind velocity of $\sim$ 6o km/s can be excluded, since within the 6 days the cloud was seen it would have moved by 45 $R_\odot$ (3 times the observed height above the limb). The density in the observed prominence was of the order of 10^{12} cm^{-3}, about a factor of 1o higher than in the surrounding chromosphere. If the excess pressure was balanced by magnetic fields, a field strength of $\sim$ 4 Gauss would have been necessary.

The empirical model chromosphere of ζ Aur also provided a test of a method developed to derive the electron density, electron temperature and the geometrical extent of giant star chromospheres from the CII (UV o.o1) λ 2325 $\overset{o}{A}$ multiplet and the CII 1335 $\overset{o}{A}$ resonance doublet (Stencel et al., 1981; Carpenter et al.,1985; Brown and Carpenter, 1985).

Using the CII method, it was claimed that coronal stars, i.e. stars below the wind/corona 'dividing line', have a small geometrical extent ($R/R_* < 1.oo1$) of the CII emitting region, while noncornal stars have typically $R/R_* \gtrsim 2$ (1.4 to 5).

Application of the CII 2325 $\overset{o}{A}$ method to a double shift high-resolution spectrum taken of ζ Aur during total eclipse of the B star - at other phases the wavelength range around 23oo $\overset{o}{A}$ is dominated by the B star - yielded the following results (Schröder et al., 1987): i) the CII 2325 $\overset{o}{A}$ flux is matched with the empirical model chromosphere obtained from eclipse data (Schröder, 1985), ii) most CII 2325 $\overset{o}{A}$ is emitted by the innermost chromosphere, iii) line intensity ratios within the CII 2325 $\overset{o}{A}$ multiplet are affected by optical depth effects, at least in supergiants, and iv) the method to determine the geometrical extent of giant star chromospheres from CII emission gives results which are quantitatively incorrect.

5. Temperatures

In case of 32 Cyg and 22 Vul, the wind electron temperature T_e has been estimated from the observed population of excitet FeII levels.

In the wind of 32 Cyg, at distances of more than 5 R_K (K supergiant radii), Che-Bohnenstengel (1984) derived a wind electron temperature $T_e \approx$ 48oo K for 1 % hydrogen ionization (n_e/n_H = o.o1) and $T_e \approx 10^4$ K for smaller electron densities.

In the chromosphere of 32 Cyg, Schröder (1986) estimated $T_e \gtrsim$ 85oo K at o.2 R_K and an increase to about 11 ooo K at o.5 R_K height. Hydrogen ionization appears to increase over the same range from about $n_e/n_H = 10^{-3}$ to 10^{-2}. This means that strong nonradiative heating occurs in heights above the photosphere where the wind starts and most of the wind driving energy - mainly potential energy, cf. Table 2 - is deposited. It can be shown that for the semiempirical velocity law as shown in Fig.12, the energy deposition into the wind per unit mass reaches its maximum at $\sim$ 1 R_K above the photosphere.

The results for 32 Cyg are consistent with radio observations of α Ori which imply an extended chromosphere with a temperature in the range 7ooo-9ooo K (Wischnewski and Wendker, 1981). In the semiempirical model of the outer atmosphere of α Ori by Hartmann and Avrett (1984), which matches line profiles of CaII, MgII, H_α and SiII, microwave emission, and a mass-loss rate of 10^{-6} $M_\odot$/yr, chromospheric temperatures of 5ooo-8ooo K extend outwards to about 1o R_*.

Similarly, Drake and Linsky (1986), cf. Linsky (1986), obtained a semiempirical model from the observed MgII k of α Boo (K2III) in which wind velocity and electron temperature rise steeply to their maximum value of $\sim$ 4o km/s and $\sim$ 8ooo K at $\sim$ 1.2 R_*, and in which there is a broad temperature plateau with $T \gtrsim$ 7ooo K extending outwards to $\sim$ 13 R_*.

In 22 Vul, a further G2Ib-II Zeta Aur type eclipsing binary, which is in several respects (high wind velocity, rotation, intermediate mass) similar to hybrid stars like α Aqr, a wind electron temperature of $3o \pm 1o \cdot 10^4$ K was estimated with the assumption of pure electron collision excitation (Reimers and Che-Bohnenstengel, 1986). However, since radiative excitation via high levels cannot be excluded at present, it is highly desirable to prove or disprove the existence of a high-temperature wind in a G supergiant. The location of 22 Vul in Fig.2 is also not inconsistent with a relatively high wind temperature.

There is little direct evidence for wind temperatures at large distances from the stars. Only in the case of α Her, from the observed absence of lines from excited fine structure levels of TiII, one can exclude that $T \gg 1oo$ at a distance of 3oo M giant radii (Reimers, 1977b), consistent with adiabatic cooling of the wind at large distances ($T \sim r^{-4/3}$).

III. The solar wind

All spacecraft observations of the solar wind have been limited to $\pm$ 1o degrees in the ecliptical plane. The following brief summary of what I consider relevant for comparison with stellar observations relies heavily on Feldman et al. (1977), Zirker (1984) and Dupree (1986).

1.) The mean solar wind has at 1 AU a wind velocity $\bar{v}_{w\odot}$ = 47o km/s, a proton particle density $\bar{N}$ = 9 cm^{-3} , and assuming spherical symmetry, a corresponding mean mass-loss rate, $\dot{M}_\odot$ = $2 \cdot 10^{-14}$ $M_\odot$ yr^{-1} .

2.) The solar wind consists of
    i) high speed streams with v_w > 65o km/s up to 9oo km/s ($\bar{v}$ = 7oo km/s, $\bar{N}$ = 4 cm^{-3})
    ii) interstream regions with v_w as low as 3oo km/s ($\bar{v} \approx$ 33o, $\bar{N} \approx$ 11).
High speed streams are from the socalled 'coronal holes'. These are large scale regions of the solar corona with T $\leq$ 1o^6 K and low density which are "dark" in X-rays. Coronal holes, magnetically open regions, are always present at the poles. Near minimum they can approach the solar equator, and in the rising part of the solar cycle coronal holes - not related to polar holes - can appear also at low heliographic latitudes.
    Between the high speed streams, the slow wind seems to arise from the belt of equatorial streamers.
    27 day averages over 6 years are shown for illustration in Fig.8.
    Solar wind variations over $\sim$3 solar rotation periods as measured at 1 AU and between 6.4 and 6.9 AU are shown in Fig.9.

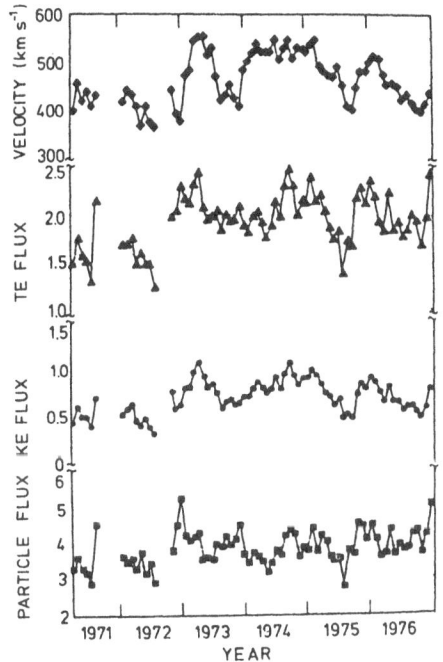

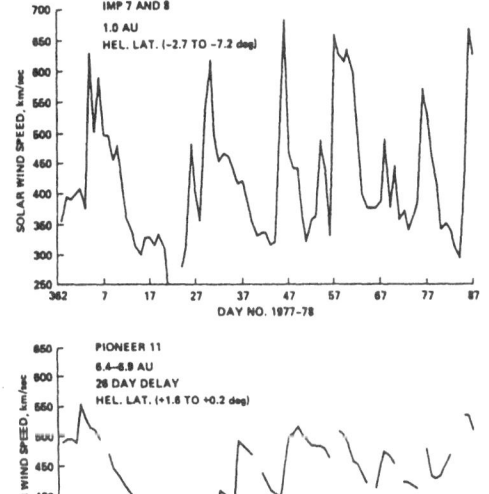

Figure 8. The 27-day averages of the solar wind proton speed, total energy flux, kinetic energy and proton number flux (from Zirker, 1984).

Figure 9. Solar wind speed for IMP 7 and 8 and Pioneer 11 for the 1978 alignment (Collard et al., 1982).

3.) Propagation of the solar wind

The propagation of the solar wind has been studied between o.3 AU and 1 AU by Helios 1 and outwards to $\sim$15 AU by plasma experiments on Pioneer 1o and 11.
- The mean solar wind speed decreases very little between 1 and 15 AU. High speed

streams and interstream regions observed near 1 AU can still be identified beyond the orbit of Jupiter (Fig.9), although the speed modulation is decreased due to stream interactions.

- The solar wind velocity fluctuations are damped with an e-folding length of 11 AU (Fig.1o), i.e. $\sigma = \sigma$ (1 AU) $\cdot$ exp. $\left((r-1)/11 \text{ AU} \right)$ (Collard et al., 1982). It has been concluded that beyond distances of 3o to 4o AU the wind may be nearly uniform and expanding radially with constant velocity.

- The flux of momentum is constant with distance from the Sun, i.e. $n \sim v^{-2}$ (if the radial variation $\sim r^{-2}$ is taken out !)

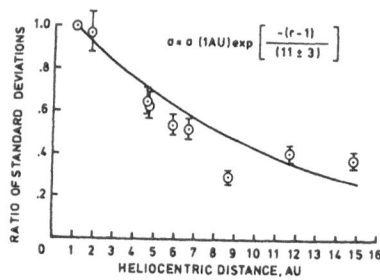

Figure 1o.   Ratio of Pioneer to IMP standard deviations of solar wind speed plotted as function of mean heliocentric distance of Pioneer spacecraft from each alignment (Collard et al., 1982).

4.) Variation of wind with solar latitude and solar cycle

All information about the solar wind at latitudes larger than $1o^{\circ}$ is from observations of interplanetary scintillations of extragalactic radio sources. Coles et al. (198o) investigated the dependence of the speed of the polar wind on the solar cycle and found a strong dependence of the latitude-dependence of the solar wind on sunspot cycle phase (cf. Fig.9 in Zirker (1984)).

On the other hand the average speed of the solar wind as observed in the ecliptic does not change much throughout the cycle. At large distances from the Sun, the proton flux density appears to be isotropic within ~ 15 % from observations of scattering of solar Ly α by interstellar neutral hydrogen. It was shown by Witt et al. (1979) that while the solar wind speed increases by ~ 45 % from the equator to the pole - consistent with interplanetary scintillation measurements - , the wind proton density decreases by $\sim 4o$ %.

Altogether, Zirker (1984) comes to the conclusion that the mass-loss of the Sun is probably isotropic in space and constant in time to $\sim 2o$ %.

5.) Spectroscopic observations of the solar wind ?

It is an interesting question to ask how small the rate of mass-loss of a late type star can be to be discovered spectroscopically. The Sun, if observed as a star, i.e. without angular resolution, shows no signs of mass-loss.

There is some (weak) evidence for outflow at coronal levels in coronal holes with velocities o (1o km/s) as observed in lines of Mg x to Fe XII (see discussion by Dupree, 1987).

Spectroscopic observations of the solar wind acceleration region using coronagraphic techniques indicate flow speeds of $\sim 1oo$ km/s at 4 $R_{\odot}$ (Withbroe et al. 1982).

IV. Stellar winds and the solar wind: Trends

1) Wind velocities

The massive stellar winds of cool stars are characterized by low asymptotic flow speeds v_w , usually less than the surface escape velocity v_{esc} . The energy required to drive the winds is $\dot{E} = \frac{1}{2} \cdot \dot{M} \cdot (v_w^2 + v_{esc}^2)$, i.e. the sum of the kinetic energy $\dot{E}_{kin}$ of the wind and the energy E_{pot} required to lift the wind out of the gravitational field of the stars. Usually the second term dominates in large, low gravity stars. The distribution of wind energy between kinetic energy and potential energy

Table 2. Compilation of data for stars with reliable wind studies (for Figs. 2 and 11)

	Stellar data					Wind data				
Star	Sp.	T_e	$\frac{R}{R_\odot}$	$\lg \frac{L}{L_\odot}$	$\frac{M}{M_\odot}$	$\dot{M}\ [M_\odot\,yr^{-1}]$	v_w [km/s]	v_{esc}	$\frac{\dot{E}_{wind}}{4\pi R^2}$ [$10^5\,erg\,cm^{-2}\,s^{-1}$]	Source for wind data
α Ori	M 2 I ab	3500	740	4.87	~20	**2** $\cdot 10^{-6}$	11	102	2	1
α Sco	M 1.5 I ab	3540	625	4.68	~18	$2.2 \cdot 10^{-6}$	17	105	1.5	2
						$1 \cdot 10^{-6}$				3
						$1.6 \cdot 10^{-6}$				4
						$0.7 \cdot 10^{-7}$				5
α¹ Her	M 5 II	3300	178	3.29	1.7	$1.1 \cdot 10^{-7}$	8	61	0.7	6
32 Cyg	K 5 I ab	3800	188	3.82	8	$2.8 \cdot 10^{-8}$	60	128	0.84	7
31 Cyg	K 4 I b	3800	202	3.91	6.2	$4 \cdot 10^{-9}$	80	109	0.96	7
ζ Aur	K 4 I b	3950	140	3.41	8.3	$6 \cdot 10^{-8}$	40	153	0.38	7
δ Sge	M 2 II	3600	140	3.43	8	$2 \cdot 10^{-8}$	28	149	1.3	8
22 Vul	G 3 Ib-II	5200	≳40	2.99	4.3	$6 \cdot 10^{-9}$ ::	160	203	≲13	9
α Boo	K 1 III p	4250	28	2.36	0.5	$6 \cdot 10^{-10}$	40	83	0.11	10
						$(< 7 \cdot 10^{-9})$			(< 3.8)	
Sun	G 2 V	5780	1	1	1	$2 \cdot 10^{-14}$	470	620	0.63	11
HR 8752	G 0 I a	5000	1000	5.5	30	10^{-6}	30	108	6.7	12
α Aqr	G 2 I b		120		~5		127	~130		13
Hybrids	K 3, 4 II		85		~4		80-180	~135		14
δ And	K 3 III		22		1.8		300	~170		15
o Cet			210		1.2		5.6	47		16
HR 3153	M 1 II		162		5		38	110		17

(::) for 30% hydrogen·ionization (upper limit is $7 \cdot 10^{-9}$ with $H^+/H = 0.01$)

(1) Mauron (1985) (2) Bowers and Knapp (1986) (3) Hagen et al. (1987) (4) Hjellming and Newell (1983)
(5) Kudritzki and Reimers.(1978) (6) Reimers (1977 b) (7) Che et al. (1983) (8) Reimers and Schröder (1983)
(9) Reimers and Che-Bohnenstengel (1986) (10) Linsky (1986) (11) Zirker (1986) (12) Lambert and Luck (1978) (13) Table 1
(14) Table 1 (15) Judge et al. (1986) (16) Reimers and Cassatella (1985) (17) Reimers (1977 c)

can be seen in more detail in Fig.2 where we plot wind velocity v_w versus escape velocity v_{esc} (data from Table 2). Virtually all cool stars are placed between the lines $v_w = v_{esc}$ and $v_w = 10^{-1} v_{esc}$ which corresponds to $\dot{E}_{kin} = \dot{E}_{pot}$ and $\dot{E}_{kin} = 10^{-2} \cdot \dot{E}_{pot}$.

In the range $v_{esc} \sim 100$ to 160 km/s stars occupy the whole range between $v_w/v_{esc} = 1$ and 10^{-1}. G supergiants (e.g. α Aqr and 22 Vul) and KII hybrid atmosphere stars are close to $v_w = v_{esc}$, M supergiants like α Ori are near $v_w = 10^{-1} v_{esc}$, while MII giants and K supergiants like 32 Cyg occupy the intermediate regime.

Although the data basis is still too small it appears that above $v_{esc} = 160$ km/s and below 80 km/s the stars are close to $v_w = v_{esc}$ and $v_w = 10^{-1} v_{esc}$, respectively. We have yet no interpretation for the transition from the "high velocity mode" (Solar type winds and hybrid star winds) to the "low velocity mode" (M giants and supergiants) where nearly all the driving energy of the wind is used to lift the escaping matter out of the potential well. The way of transition from one type to the other (Fig.2) may be an indication for a transition from one dominating wind acceleration mechanism to another mechanism while in the intermediate regime both (or several) mechanisms operate simultaneously.

2) Mass-loss rates and driving energies

Further quantities that can teach us something about the driving mechanism(s) of the winds are the total energy requirements for winds and the fraction of the stellar luminosity that goes into the mass-loss process.

If always the same fraction of the stellar luminosity is used to drive the mass-loss, and the kinetic energy of the wind is small compared to the potential energy, the mass-loss rate scales as $\dot{M} \sim L/g \cdot R$. It was shown (Reimers, 1975) that for Pop.I red giants this is a reasonable approximation formula. With the more accurate mass-loss rates available now we can look more closely into this question.

Fig.11 shows that $\dot{M} = 5 \cdot 10^{-13} L/g \cdot R$ is indeed a good approximation for most of our stars of Table 2. Within a factor of 2 (not for all stars is $\dot{E}_{kin} \gg \dot{E}_{pot}$) this corresponds to $\dot{E}_w \sim 2.5 \cdot 10^{-5} \cdot L$. All stars seem to lie above $\dot{E}_w/L = 10^{-6}$ (see also Table 2). As can be seen from Table 2, the energy requirements (wind energy flux per cm^2 stellar surface) even for supergiants with mass-loss rates 10^8 times that of the Sun are rather modest. Most cool stars require about 10^5 erg cm^{-1} s^{-1} although the total observed range covers a factor of ~ 40, and the G supergiants show the highest values (10^6 erg cm^{-2} s^{-1}).

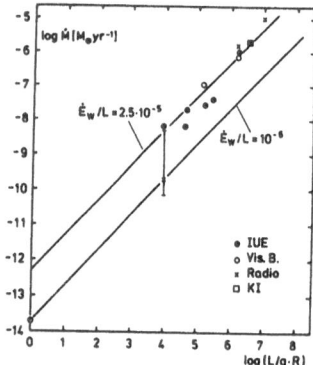

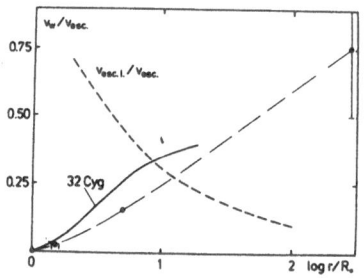

Figure 11. Mass-loss rates from Table 2 versus $L/g \cdot R$ (solar units). Star with the large error bar is α Boo (upper limit: 1 % H$^+$, lower limit: 100 % H$^+$).

Figure 12. Semiempirical wind acceleration law $v(r)$ for 32 Cyg and the Sun (o) normalized to $v_{esc}^2 = 2GM/R$. Also plotted is the normalized local escape velocity $v_{esc.l.}^2 = 2GM/r$.

Compared to chromomospheric energy losses (represented by MgII and CaII), the wind energy losses are usually small (Table 3), except perhaps in the latest, most luminous stars. It can also be seen that while transition layer and coronal emission decrease steeply with decreasing gravity of the stars (factor of $> 10^3$ between Sun and α Ori), a corresponding variation of wind energies is not observed. On the other hand, the comparison of the G supergiant β Dra, which has very high chromospheric and coronal activity and no observable wind (closed magnetic structures ?), with 22 Vul which has nearly the same spectral type and a strong wind shows that at a given location in the HR diagram there may be a considerable energy range used for a wind. This is - in a way - similar to what is known from chromospheric CaII and MgII fluxes which vary by a factor ~ 10 at a given location in the HR diagram. A relation between chromospheric energy losses and wind energy losses is not known.

2) Wind acceleration and wind 'turbulence'

a) Wind acceleration

Empirical models of the wind acceleration regions in the Sun and in cool giants could be useful for identifying mass-loss mechanisms. In stars, the combination of empirical chromospheric models up to heights of ~ 1 R_{*} with wind data in ζ Aur type eclipsing binaries allowed to construct a semiempirical velocity law $v(r)$ (sect. II), while for the Sun only two crude determinations of coronal expansion at 0.5 $R_\odot$ and 4 $R_\odot$ are available (see sect. III).

A comparison is made between the K supergiant 32 Cyg and the Sun in Fig.12. Apparently the wind velocity increases more steeply close to the star in the giant. This is consistent with the fact that for 32 Cyg the driving energy of the wind is mainly potential energy (cf. Fig.2) which also must be fed into the wind close to the star. Both in the Sun and in 32 Cyg, the wind velocity surpasses the local velocity of escape at ~ 10 R_{*} . In giants like 32 Cyg, the relatively low asymptotic flow speed ($v_\infty/v_{esc} = 0.4$) constrains the amount of energy added to the flow in the supersonic regime (cf. discussion by Holzer and McGregor, 1985).

b) Wind 'turbulence'

The only quantitative measure of the dispersion of wind velocities in a star is 'microturbulence' v_t as determined from circumstellar lines. Observed values in red giants are typically half the wind velocity, $v_t \approx \frac{1}{2} v_w$ (Reimers, 1981, 1987). This is not very different from what we see in the solar wind. The solar wind velocity variations (300 - 900 km/s) observed at 1 AU would also mimic a large microturbulence if the solar wind was detectable spectroscopically.

There is also direct observational evidence that at large distances from the stars microturbulence has declined, like in the Sun, where the velocity dispersion is damped with distance from the Sun with an e-folding length of 11 AU.

In α Ori, e.g., the main S1 component of the circumstellar CO 1-o vibration rotation band at 4.6 μ (seen also in KI and NaI) yields $v_t = 4$ km/s and $v_w = 11$ km/s, while the weaker S1 component, probably formed at larger distances from the star, has $v_t = 1$ km/s and $v_w = 18$ km/s (Bernat et al. 1980).

The quantitative analysis of thermal mm-wave CO lines, observed in far evolved late type stars at large distances from the stars, also shows that for $v_w = 20$ km/s typically $v_t \approx 1 \ldots 2$ km/s is required (K. Schönberg, priv.comm.).

Table 3. Energy losses $\left(10^5 \text{ erg s}^{-1}\right)$ per cm^2 of the stellar surface in various cool stars

		CaII H+K	MgII	CIV,SiIII, NV	Soft X-rays	$\dot{E}_{wind}$	$\dfrac{\dot{E}_{wpot}}{\dot{E}_{wkin}}$	$\dfrac{\dot{E}_{wind}}{L}$ $\left(10^{-5}\right)$
Quiet Sun	G2V	12	9	0.09	0.1	0.63	1.7	0.1
α Boo	K1IIIp	7.5	1.8	<0.003	<0.0007	0.11 (<3.8)	4.5	0.06 (<2.1)
α Aqr	G2Ib	30	50	0.34	<0.0001	-	-	-
22 Vul	G3Ib-II	-	-	-	-	13	1.6	3.2
β Dra	G2II-Ib	41	28	0.85	1.2	no wind seen		
α Ori	M2Iab	-	0.26	<0.0001	<0.0001	2	90	2.4

4. Concluding remarks

According to the trends and numbers which we have presented in Figs. 2 and 11 and Tables 2 + 3, the solar wind seems to behave quite 'normal' among winds of other late type stars. We should remember, however, how little we know about stellar winds, how few reliable numbers are available. Cool stars' winds are probably as non-steady and non-spherically symmetric as the solar wind. Most probably more than one mass-loss mechanism is operative in late type stars: The solar wind is driven by the hot corona and is further accelerated by magnetic waves, and in the stars of intermediate escape velocities ($100 \lesssim v_{esc} \lesssim 160$) there is evidence for a transition from one dominating mechanism to another (thermally driven winds $\rightarrow$ wave driven winds ?).

In addition, the influence of stellar activity (magnetic fields) on winds in giant stars has not yet been explored. Hybrid atmosphere giants do show wind variations and variations of chromospheric UV emission lines so that a detailed, long-term study with a future UV satellite is promising.

It will also be important to try to explore winds in G and K giants of somewhat lower luminosity (higher v_{esc}), e.g. with the binary technique. One can estimate that in the MgII lines mass-loss at rates at least as low as 10^{-11} is detectable. The scarce of data in Figs. 2 and 11 shows that we are certainly only at the beginning of a study of stellar winds in the context of the solar wind.

References

Baade,R. 1986, Astron.Astrophys. 154, 145
Bernat,A.P., Hall,D.N.B., Hinkle,K.H., Ridgeway,S.T. 1979, Astrophys.J. 233, L135
Bowers,P.F., Knapp,G.R. 1986 in "Workshop on the Late Stages of Stellar Evolution", Calgary
Brown,A. 1986, in Cool Stars, Stellar Systems, and the Sun (M. Zeilik, D.M. Gibson eds.) Lecture Notes in Physics, Springer, Berlin etc. p. 454
Brown,A., Carpenter,K.G. 1984, Astrophys.J. 287, L43
Brown,A., Reimers,D., Linsky,J.L. 1986 in New Insights in Astrophysics, Proc. Joint NASA/ESA/SERC Conference p. 169
Cacciari,C., Freeman,K.C. 1983, Astrophys.J. 268, 185
Carpenter,K.G., Brown,A., Stencel,R.E. 1985, Astrophys.J. 289, 676
Che-Bohnenstengel,A., 1984, Astron.Astrophys. 138, 333
Che-Bohnenstengel,A., Reimers,D. 1986, Astron.Astrophys. 156, 172
Che,A., Hempe,K., Reimers,D. 1985, Astron.Astrophys. 126, 225
Cohen,J. 1976, Astrophys.J. 1o3, L127
Coles,W.A. et al. 1980, Nature 286, 239

Collard,H.R., Mihalow,J.D. Wolfe,D.H. 1982, Journ.Geophys.Rev. 87, 22o3
Deutsch,A.J. 1956, Astrophys.J. 123, 21o
Drake,S.A. 1986 priv. comm.
Drake,S.A., Linsky,J.L. 1986, Astron.J. 91, 6o2
Drake,S.A., Brown,A., Linsky,J.L. 1984, Astrophys.J. 284, 774
Dupree,A.K. 1986, Ann.Rev.Astron.Ap. 24, 377
Dupree,A.K., Reimers,D. 1987, in: The Scientific Accomplishments of the IUE,
    Reidel, Dordrecht/Holland
Dupree,A.K., Hartmann,L., Avrett,E.H. 1984, Ap.J. 281,L37
Feldmann,W.S. et al. 1978, Journ.Geophys.Rev. 83, 2177
Gahm,G., Hultquist,L. 1972, Astron.Astrophys. 16, 329
Goldberg,L. 1979, Q.J.R.A.S. 2o, 361
Goldberg,L. Hege,E.K., Hubbard,E.N., Strittmatter,P.A. Cocke,W.J. 1981, SAO Spc.Rep.
    392, 131
Hagen,J.-J., Hempe,K., Reimers,D. 1987, Astron.Astrophys. in press
Haisch,B.M., Simon,T. 1982, Astrophys.J. 263, 252
Hartmann,L., Dupree,A.K., Raymond,J.C. 198o, Astrophys.J. 236, L143
                                      1981, Astrophys.J. 246, 193
                                      1982, Astrophys.J. 252, 214
Hartmann,L., Avrett,E.H. 1984, Astrophys.J. 284, 238
Hege,E.K., Hebden,J.C., Christou,J.C. 1986 in Cool Stars, Stellar Systems, and the
    Sun (M. Heilik, D.M. Gibson eds.), Springer Verlag Berlin etc. p.414
Hempe,K. 1982, Astron.Astrophys. 115, 133
Hempe,K. 1984, Astron.Astrophys.Suppl. 56, 115
Hempe,K., Reimers,D. 1982, Astron.Astrophys. 1o7, 36
Hjellming,R.M., Newell,R.T. 1983, Astrophys.J. 275, 7o4
Holzer ,T.E., MacGregor,K.B. 1985, in Mass Loss from Red Giants
    (M. Morris, B. Zuckermann eds.) D. Reidel, p. 229
Judge,P.G., Jordan,C., Rowan-Robinson,M. 1987, M.N.R.A.S. 224, 93
Kraft,R.P., Preston,G.W., Woolf,S.C. 1964, Astrophys.J. 14o, 235
Kudritzki,R.P., Reimers,D. 1978, Astron.Astrophys. 7o, 227
Lambert,D.L., Luck,R.E. 1978, M.N.R.A.S. 184, 4o5
Linsky,J.L. 1981, Ann.Rev.Astron.Ap. 18, 439
Linsky,J.L. 1986, Irish Astron.J. 17, 343
Linsky,J.L., Haisch,B.M. 1979, Astrophys.J. 229, L27
Mauron,N. 1985, Doct.Thesis, Univ. Toulouse
Mauron,N., Fort,B., Querci,F., Dreux,M., Fauconnier,T., Lamy,P. 1984, Astron.
    Astrophys. 13o, 341
Mauron,N., Cailloux,M., Tilloles,P., Lefèvre,O. 1986, Astron.Astrophys. 165, L9
Middelkoop,F. 1982, Astron.Astrophys. 113, 1
O'Brien,G.T., 198o, Ph.D. Thesis, Univ. Texas, Austin
O'Brien,G.P., Lambert,D.L. 1986, Astrophys.J.Suppl 62, 899
Oznovich,I., Gibson,D.M. 1986, in Cool Stars, Stellar Systems, and the Sun (M. Zeilik,
    D.M. Gibson, Eds.), Springer-Verlag, Berlin etc. p. 124
Reimers,D. 1975a, Mem.Soc.Roy.Sci Liège 6e Ser. 8, 369
            1975b, in Problems in Stellar Atmospheres and Envelopes (eds. B. Baschek,
                W.H. Kegel, G. Traving), Springer-Verlag, Berlin etc. p. 229
            1977a, Astron.Astrophys. 57, 395
            1977b, Astron.Astrophys. 61, 217 (Erratum 67, 161)
            1977c, Astron.Astrophys. 54, 485
            1981, in Physical Processes in Red Giants (I. Iben, A. Renzini eds.)
                D. Reidel, p. 269
            1982, Astron.Astrophys. 1o7, 292
            1984, Astron.Astrophys. 136, L5
            1987, in Circumstellar Matter (IAU Symp. 122) D. Reidel, in press
Reimers,D., Cassatella,A. 1985, Astrophys.J. 297, 275
Reimers,D., Che-Bohnenstengel,A. 1986, Astron.Astrophys. 166, 252
Reimers,D., Schröder,K.P. 1983, Astron.Astrophys. 124, 241
Robel,E. 1987, Diplomarbeit Univ. Hamburg
Schröder,K.-P. 1983, Astron.Astrophys. 124, L16
              1985a, Astron.Astrophys. 147, 1o3
              1985b, Doct.Thesis, Univ. Hamburg

Schröder,K.-P.,1986, Astron.Astrophys. 17o, 7o

Schröder,K.-P., Reimers,D., Carpenter,K.G., Brown,A.,1987, Astron.Astrophys. submitted

Smolinski,J., Feldmann,P.A., Higgs,L.A. 1977, Astron.Astrophys. 6o, 277

Stencel,R.E., 1978, Astrophys.J. 223, L37

Stencel,R.E., Mullan,D.J. 198o, Astrophys.J. 238, 221

Stencel,R.E., Linsky,J.L., Brown,A., Jordan,C., Carpenter,K.G., Wing,R.F., Czyzak,S.
   1981, M.N.R.A.S. 196, 47

Wilson,O.C. 196oa, Astrophys.J. 132, 136

           196ob, in Stellar Atmospheres (ed. J.L. Greenstein), Univ. Chicago Press,
               Chicago

Wischnewski,E., Wendker,H.J. 1981, Astron.Astrophys. 96, 1o2

Witt,N., Ajello,J., Blum,P. 1979, Astron.Astrophys. 73, 272

Withbroe,G.L., Kohl,J.L., Weiser,H., Munro,R.H. 1982, Space Sci.Rev. 33, 17

Zirin,H. 1976, Astrophys.J. 2o8, 414

Zirker,J.B. 1984, in Effects of Variable Mass Loss on the Local Stellar Environment
   (R. Stalio, R.N. Thomas eds.), Trieste, p.25

IV. OBSERVATIONS FROM SPACE

OBSERVATIONS FROM SPACE VS. GROUND BASED OBSERVATIONS: ADVANTAGES AND DISADVANTAGES

F. Kneer

Universitäts-Sternwarte

Geismarlandstr. 11, D-3400 Göttingen

ABSTRACT

A comparison of observations from space and from the ground should be limited to the optical spectral range (about 0.35µ-10µ). Optical observations give, either directly or indirectly, access to:

a) the photosphere and chromosphere where most research is into small-scale dynamical processes, and

b) the solar interior by means of seismology, measurement of large-scale velocities, and statistical analysis of long-period data sets.

Instrumentation and observing modes have to comply with the scientific needs. With the aid of a few examples we discuss advantages and disadvantages of observations from space and from the ground.

a) Observations with high spatial resolution are limited, principally, by the short time-scales (10s - 100s) of the dynamic phenomena and by photon noise. An observational cot (e.g. a time sequence) is thus restricted to few spectral bands and to a small field of view. Criteria for the selection of the instrumentation are: spatial resolution, duration of observation, and versatility of the instrument.

b) For "observing" the solar interior one needs velocity measurements of high precision (1 m s^{-1} - 1 cm s^{-1}) and over extended intervals of time (years to centuries) on a routine basis. Selection criteria to decide on space or ground based observations are stability and the possibility to monitor the Sun over years.

1. INTRODUCTION

Very soon after World War II the space science era began with the use of V2 missiles for research into solar physics and for the exploration of the ultraviolet spectrum of the Sun, which is not accessible from the ground. There is no doubt: Since then a wealth of information has been gathered which one could not have imagined before.

By means of UV spectroscopy and in situ particle measurements fundamental insight has been gained on the physics of the outer solar atmosphere and its relation to the Earth's environment.

Wy do we have to go into space for observing the Sun in the visible spectral range as well, which is accessible from the ground? Observation with high spatial and temporal resolution revealed the importance of small-scale dynamics for stellar atmospheric structure. It is, however, recognised that high resolution observations from the ground are strongly hampered, if not made impossible most of the time, by the turbulence of the Earth's atmosphere, i.e. by bad seeing. From here came the main thrust into solar space observation and led to balloon borne telescopes as early as 1956. Indeed, efforts have already been undertaken three decades ago to avoid image degradation by the Earth's atmosphere. Table 1 gives the data of the solar instruments for the visible spectral range which have been flown so far. (I omit the Hα auxiliary telescopes for UV observations as well as the coronagraphs aboard ATM and SMM). The Soviet Stratospheric Observatory was described to have a spatial resolution corresponding to a 50cm telescope. Excellent time sequences on granular dynamics were obtained with Spektrostratoskop and the US instrument aboard Spacelab 2, the latter being a predecessor to a larger, highresolution telescopeand the only one being truly a space telescope among the "stratospheric" observatories.

Table 1

Instrument	(cm)	flights	Reference
manned balloon flight	29	1956	Blackwell et al.,1959
Stratoscope	30	1957 1959(4x)	Danielson, 1961
Soviet Solar Stratospheric Observatory	100	1966, 1968 1970, 1973	Krat et al., 1970 Krat, 1981
Spektrostratoskop	30	1975	Mehltretter, 1978
SOUP	30	1985	Title, 1985

Certainly, during the last decades, ground-based facilities made a large progress. Earth-bound observatories on excellent sites and with sophisticated telescope constructions designed to minimize both internal and external seeing have proven to be scientifically very valuable, and there is hope that one can recover the genuine information from the Sun using speckle techniques and modern computers. Thus, today as before space borne and ground-based observatories are competitive in the sense that, with both "observing modes", we aim at highest resolution obtainable. Yet competition is only one aspect. The other aspect is that the two modes are complementary in the sense that in many applications one technology allows to solve problems which may not be attacked by the other. This will also hold during the foreseeable future.

In what follows, starting from some facts about our object, the Sun, and by means of a few examples, I should like to throw open to discussion a few points of view on how to judge upon advantages and disadvantages of observations from space and from ground.

2. THE SUN IN THE OPTICAL SPECTRAL RANGE

Figure 1 gives a synopsis of the wavelength dependence of the solar brightness temperature and the height of continuum optical depth $\tau_c = 1$ (redrawn from Vernazza et al., 1976) together with the solar radiative energy output and the transmission of the Earth's atmosphere. Instead for the energy we shall ask, below, for the number of photons emitted.

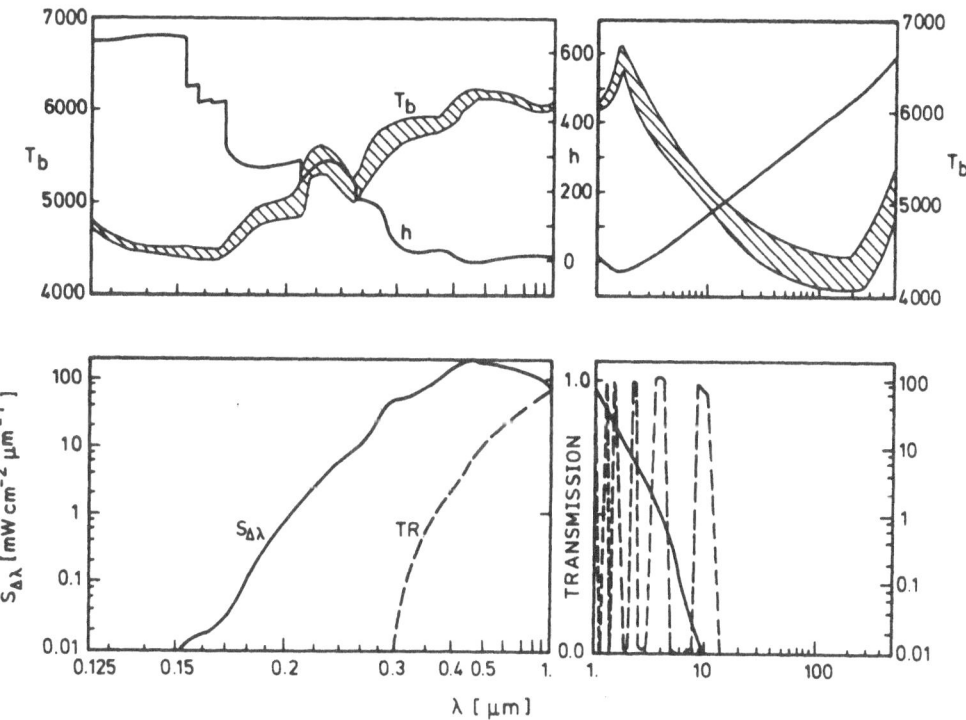

Figure 1: Solar brightness temperature T, height h at continuum optical depth $\tau_c = 1$ (from Vernazza et al., 1976), solar output $S_{\Delta\lambda}$ (Pierce and Allen, 1977, Heath and Thekaekara, 1977), and transmission TR of the Earth's atmosphere (Lamla, 1982, Plass and Yates, 1965)

Then, observing at longer wavelengths is more favorable, the maximum of emitted photons per wavelength and unit time being at a wavelength of about 1.2μ. However, if one wishes to keep high spatial resolution with a given telescope aperture one must not observe at too long wavelengths, longer than, say, 1μ.

There are three aspects of the solar layers accessible in visible light. First, the photosphere where most of the energy is emitted forms the skin between the stellar body and the outside space. We cannot see deeper (apart from measuring the gravitational potential and the neutrino flux). To investigate the interior one has to use the information shown on the surface. Secondly, this skin which receives energy and mass from below is far from equilibrium. It is in permanent motion and shows many phenomena of spontaneous organisation of structures such as granulation, waves, and magnetic fields. And third, every process occuring in the high coronal layers or in the solar wind is rooted in the photosphere and is linked through the photosphere to the interior.

By means of spectral lines one can probe the accessible part of the atmosphere on a height range of a few thousand kilometers. These are the layers where the temperature is lowest and the pressure scale height H is the smallest, about 100 km. Why is the scale height a measure so often quoted as a limit on observational needs? From radiation transfer there is the argument that the radiation field is smoothed out over this distance and changes of the state of the gas within this distance are principally hidden. From MHD considerations one realises that a gas parcel of linear dimensions smaller than the scale height has neighbouring elements with physical conditions very much the same as its own.

The scale height leads to a characteristic dynamic time, $H/C_s \approx 10s$, where C_s is the sound speed. Some processes can be even faster, e.g. the radiative relaxation, or when magnetic fields are involved. Such time scales are very restricting constraints for high resolution observations. It means that an observation, or, more precisely, one time step in a sequential observation, has to be completed within a few seconds.

Given the above restriction let us turn to another constraint, the photon flux. Its estimate will show that there are principal limits. At 5500Å the continuum intensity at disc centre is (Gingerich et al., 1971)

$$I_\nu \approx 3.6 \times 10^{-5} \text{ ergs cm}^{-2}\text{s}^{-1}\text{Hz}^{-1}\text{ster}^{-1}$$

corresponding to a photon flux of

$$N \approx 10^{18} \text{ photons cm}^{-2}\text{s}^{-1} \text{ Å}^{-1} \text{ ster}^{-1}$$

At 1 AU distance, the number of photons collected by a 1.2m telescope from a resolution element on the Sun's surface (80 km) is

$$N_{1AU} \approx 5.6 \times 10^9 \times T \times Q \text{ photons s}^{-1} \text{ Å}^{-1} \text{ (res. el.)}^{-1}$$

with T = transmission coefficient (of Earth's atmosphere, optics, filter) and Q =
detective quantum efficiency. The latter is independent of the telescope aperture
or of the distance of the telescope from the Sun. Increasing the aperture or going
closer to the Sun decreases the area on the Sun corresponding to the angular resolution.

Now, let me consider the case of a wavelength scanning spectrometer like a tunable
smallband filter on a two-dimensional field of view. To scan a wavelength range of
1Å within 1 sec (10m Å in 10 msec) one gets per measurement in a line with 10 percent
residual intensity and with an optimistic estimate of T x Q = 0.1

$$N_{mes} \approx 5.6 \times 10^3 \text{ photons } (10m\text{Å})^{-1} (10 \text{ msec})^{-1} (\text{res.el.})^{-1}$$

or a signal to noise ratio

$$s/n \approx 75.$$

The photon counts become even less favourable when observing fainter features, e.g.
the solar limb, sunspots, or prominences. Also we have not considered any oversamp-
ling or the needs for polarimetry which necessarily reduce the number of detectable
photons per measurement by a factor of 2 to 10.

This demonstrates that limits to the observable wavelength range and/or the observable
area on the Sun are given by the photon statistics when one aims at observing with
highest possible spatial resolution and with time steps comparable to or smaller
than the solar time scales. One may, of course, sacrifice spatial resolution, i.e.
build a very large photon collector and use it at lower than diffraction limited
resolution.

3. SMALL-SCALE DYNAMICS

There is no need to review the small-scale dynamics of the solar and of stellar atmospheres. A number of conferences on this subject have taken place in the last few
years (see e.g. Stenflo, 1983, Keil, 1984, Deinzer et al., 1986). Yet a few examples
may elucidate some criteria for the selection of the "observing mode": observation
from space or from the ground.

3.1 Granulation

In a recent investigation Roudier and Muller (1986) have shown that granulation
changes properties at the size of about 1.4 arcsec in the sense that the smaller
granules show an indication of turbulent cascading. Figure 2 shows a spectrogram of
the disc centre near 6500Å taken with the new Gregory-Coudé-Telescope at the Observatorio del Teide (Wiehr, 1987).

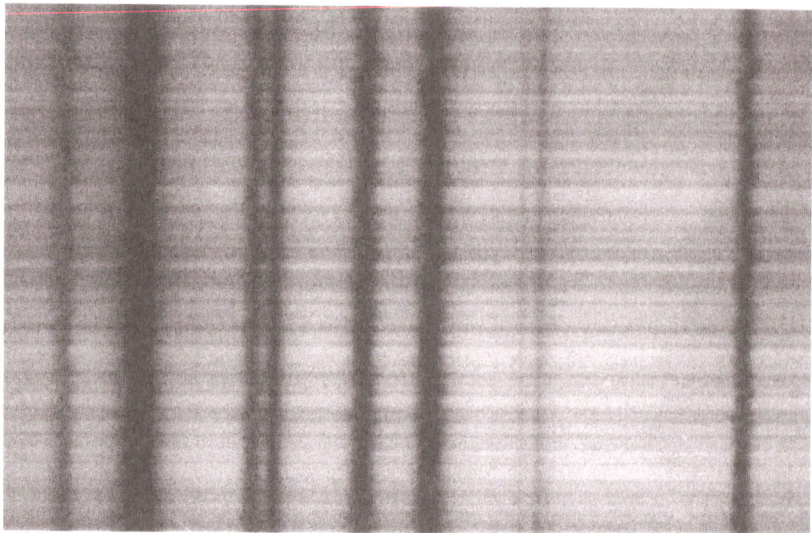

Figure 2: Photographic spectrogram from solar disc centre, 6493-6500Å, 10m echelle spectrograph of the 45cm Gregory telescope at the Observatorio del Teide, Tenerife (courtesy E. Wiehr, 1987).

The analysis of this high-resolution spectrogram lends support to the result of Roudier and Muller. Figure 3 gives coherence and phase between continuum intensity and the shift of the weak Fe I 6494Å line (from Kneer and Wiehr, 1987). The coherence breaks down at structural sizes near 1.4 arcsec and the phase becomes unstable there. In accordance with earlier findings of Durrant and Nesis (1982), line shifts and intensities loose correlation at small scales, i.e. the motion is turbulent. This breakdown is not due to instrumental noise but of solar origin, as may be learned from the fact that coherence between other spectral features on the same spectrogram extends to much smaller scales (about 0.7 arcsec, not shown here).

There would certainly be more to say about granulation, granular overshoot and secondary motions, and at other occasions during this conference there is deeper and broader discussion. Here I must be satisfied with the notion that observing the small-scale end of the granulation phenomenon may lead to understanding turbulence in stratified media.

3.2 Oscillations and waves

Figure 4 shows the contour plot of a power analysis of a Hα filtergram time sequence (Kneer and von Uexküll, 1985). The duration of the observation was 128 min. The point to be demonstrated here is that the 5-min p modes are present at high wavenumbers

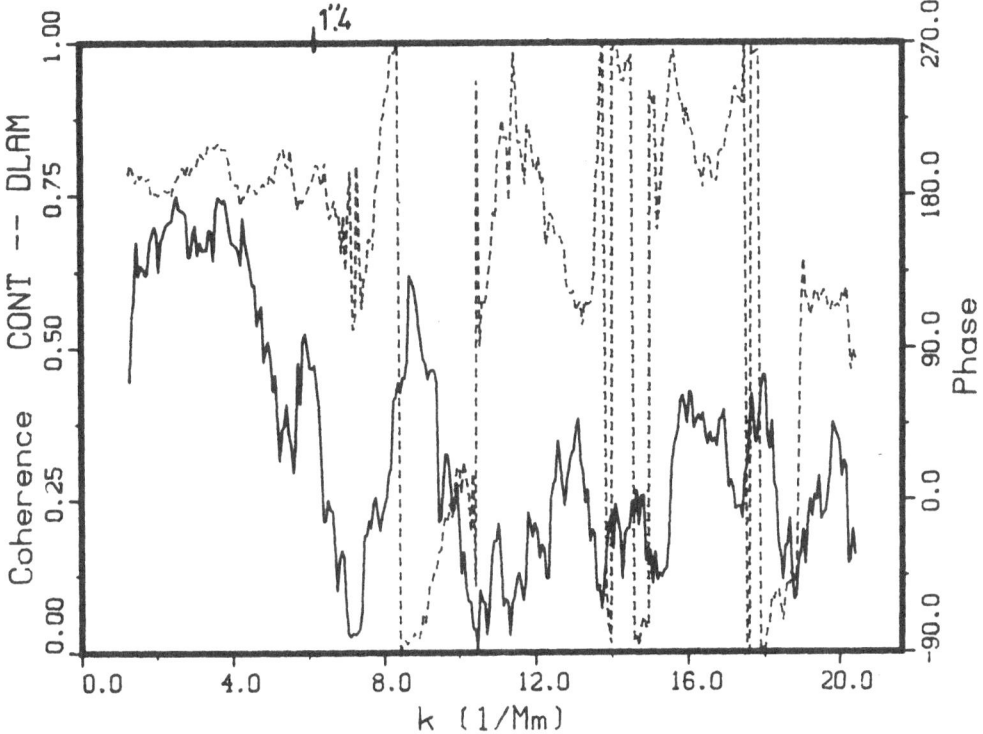

Figure 3: Coherence (full line) and phase (dashed) between continuum and shift of weak Fe I line at 6494Å, number of degrees of freedom is 42.

corresponding to high degrees (l = 2000 and beyond). The hatched area in the k- ω-diagram where Ando and Osaki (1977) find excitation by the ϰ - mechanism does by far not extend to such high degrees. Possibly, then, the waves which we see here are produced by non-linearities of the convective motion (Goldreich and Keeley, 1977, Antia et al., 1984). The resonant cavity of the modes extends roughly as 1/k into the subphotospheric layers (Gough, 1980). Thus, the presence of these modes, their frequencies and their breadth allow to probe the close subsurface layers and the amplitude of the inhomogeneities there (Bogdan and Zweibel, 1985, Nesis, 1986). The propagation properties of the high l modes are not well known (Kneer and von Uexküll, 1985), thus, it is also not clear whether they provide the high photospheric layers with non-radiative energy.

Waves with shorter periods (20s - 200s) are certainly present in the solar atmosphere (Deubner, 1976, Mein and Schmieder, 1981, von Uexküll et al., 1985, Staiger, 1987). Yet limited resolution makes an estimate of the mechanical flux and of its divergence very difficult.

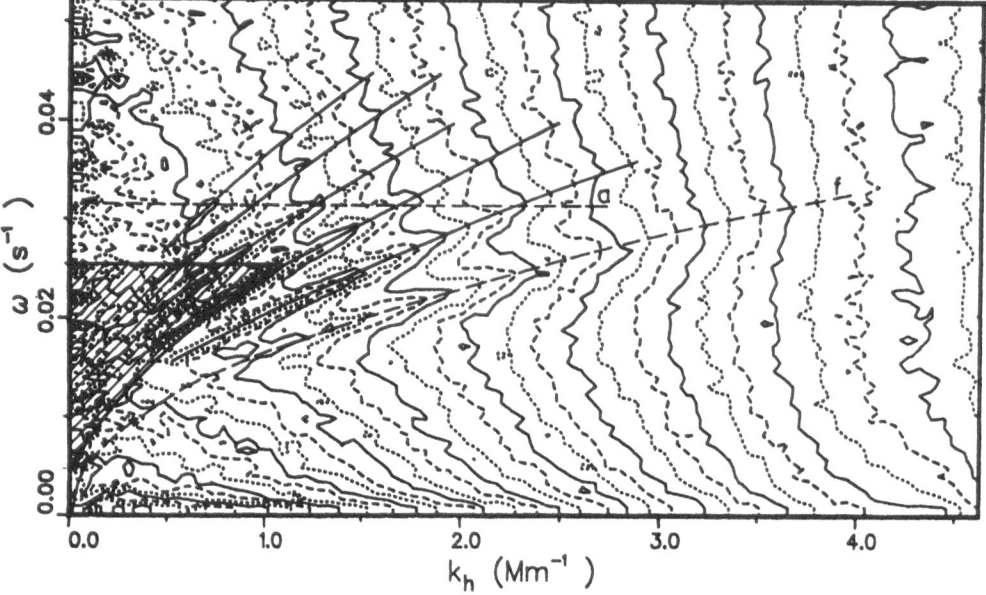

Figure 4: Power in the k-ω-plane obtained from a Hα filtergram time sequence (Kneer and von Uexküll, 1985).

3.3 Small-scale magnetic flux tubes

Much of today's research activity is focussed on this subject. And equally much speculation is possible on flux tube dynamics simply because crucial observations of high spatial resolution are missing. It is not known whether the facular points coincide spatially with magnetic field concentrations (Muller, 1985). It is not known whether the 1 - 2 k Gauss fields merge already in concentrated form the deep convective layers or whether they develop from weak fields which collapse in a con-vectively unstable layer (Parker, 1978, Hasan, 1984). The zero-crossing of the Stokes V profile shows Doppler shifts corresponding to velocities of 1 km s^{-1} when small areas are observed with sufficient temporal resolution (Scholiers and Wiehr, 1985). However, the Stokes V profile is unshifted when one averages over large areas (Stenflo et al., 1984). Even more contradictory, both the spatially non-resolved and the better resolved V profiles are asymmetric which is commonly interpreted as an effect of velocity gradients (Pahlke and Solanki, 1986, but see also Landi degl'Innicenti, 1985). Overstable oscillations of the fluxtubes, predicted by Hasan (1984) could possibly explain the observational puzzle. Resolved observations are highly needed.

There is accord and observational evidence (Skumanich et al., 1975) that small-scale
magnetic fields are related with supergranular downflow and with the chromospheric
network. Yet it is unproven that chromospheric Hα structure outlines the magnetic
field, as is commonly believed. Likewise, an excess wave flux in chromospheric network, guided by the magnetic tubes, could not yet be found. Again, observations with
high spatial and temporal resolution are needed to actually <u>see</u> the energy dissipation
related to magnetic field dynamics.

3.4 Criteria

We may now, on the basis of the foregoing few examples, discuss some guidelines for
the comparison of space borne or ground based instrumentation. High spatial resolution will be the first criterium. Furthermore, it will be agreed that, to study
dynamical, non-stationary phenomena, time sequences are needed. (Admittedly, much
insight into fine structure has been gained in the past from single snapshots. But
we could not even guess on dynamic behaviour if we would not know about the temporal
evolution.) Thus, the achievable time resolution and the duration of the observing
time will be of importance as well. And finally, the above examples on small-scale
dynamics show that the scientific demands on equipment are very diverse. The versatility may then be another criterion.

3.4.1 Spatial resolution

Possibly, the very next space facility will be the 1m High Resolution Solar Observatory (HRSO) which will resolve approximately 0.1 arcsec. Experience with its predecessor (SOT) has tought that larger solar telescopes may not be transported to
space within the near future, due to the large costs. Earth-based telescopes, although
possibly of larger aperture - the Large Earth-based Solar Telescope (LEST) is planned
with a 2.40 m primary -, are limited by seeing effects most of the time. The question
is whether seeing puts a principal resolution limit to ground-based telescopes. The
answer is negative, so far. The best (short-exposure) pictures of granulation - and
thus pioneering revelations about granulation stem from the ground (Muller, 1985,
Scharmer, 1987). Speckle techniques allow to restore the images to the point where
they are diffraction limited (von der Lühe, 1987). However, the very best seeing
conditions last only few seconds, minutes perhaps. Solar image restoration has to
struggle with tremendous amounts of data. I summarize, space telescopes give con-
stantly high resolution, while, at least to my expectation for the next two decades,
ground based telescopes give the highest resolution only at very few moments and
hinge to the application of image restoration.

3.4.2 Temporal resolution

As explained above, high temporal resolution sets demands on the photon throughput. Space telescopes have the advantage that no atmospheric attenuation is present nor vacuum windows are needed. For given aperture the throughput is thus somewhat higher than for ground telescopes. However, making the telescope on ground larger and sacrificing resolution will give the same photon counts as from a space telescope.

3.4.3 Duration of observation

Even for small-scale dynamics it may be desirable to have observations uninterrupted for several hours. Earth satellites with orbits of low inclination do not allow this. There are necessarily night gaps. Thus a polar, full sunlit orbit would be much preferable for HRSO. In this case and together with its permanently high spatial resolution HRSO will be superior to any existing solar telescope.

3.4.4 Versatility

All space telescopes flown or planned so far have a foreseen limited duration of flight and thus a very restricted observing schedule planned long time ahead. Spectrographic settings have to remain fixed to what one thinks are the most important spectroscopic features and the best choice. Observing time is limited and so are, at the end, the number of photons and the number of spectral lines to be observed. It will not be possible to repeat measurements on a later observing run with somewhat different settings, simply because followup space missions are not feasible, at least not for HRSO which is not a permanent or repeatedly flown facility. For the same reason, repair, replacement and improvement of instrumentation is not an option for HRSO.

The possibility to select various filter tunings and spectrographic settings for specific scientific questions is a desirable property of observations and is most easily obtained on the ground.

3.4.5 Phenomena to be investigated

It may well happen that during the next short space missions very little activity occurs on the Sun, e.g. that neither sunspots nor prominences are visible. Furthermore, many of the small-scale features bahave solar cycle dependent. It is thus obvious that, as long as no permanent space observatory is available, ground based observations are indispensable.

Let me compare then the expected performance of space and ground telescopes in view of the three scientific problems described above. A space telescope in an equatorial orbit will certainly bring exciting observations about granular dynamics and magnetic

flux tubes, but the frequency resolution will be insufficient to exploit the 5-min
p modes for subphotospheric investigation. In a full sunlit orbit the results on
"long lived" phenomena would be much superior and the detection of fluxtube oscillations and the investigation of their birth and decay much easier. I do not expect that observations from ground can ever be of the same constant quality. Some
may show very high resolution, single granulation pictures perhaps which may lead
the way to new detections. Some may be of medium, yet satisfactory quality to investigate p-modes and other short-period waves with various instrumental settings.
The observation of the dynamics of the small-scale magnetic fields with time-scales
of about 1/2h will be very difficult from the ground because of the deleterious
properties of atmospheric turbulence.

4. THE INTERIOR OF THE SUN

People have observed the Sun for several centuries, even for millenia (see Wittmann
and Xu, 1987a,b, and the references there). Long-term records of solar activity form
the basis for investigating stellar dynamos. Archive data obtained on a routine basis
for long-term programs, e.g. the filtergrams from the Catania Observatory, the whole
disc magnetograms from Kitt Peak, and the Mt. Wilson Dopplergrams, cover a time span
of several decades. They are used today to search for large scale motions, meridional
flows and giant cells, and to measure the solar rotation velocity and its cycle
dependence (see e.g. Howard, 1985, and the review by Schröter, 1985). Doubtless, the
data are extremely valuable for studies into the hydrodynamics of the solar interior
and efforts to continue their acquisition are worthwhile. To my knowledge, attempts
to operate competitive instruments in space over one or more solar cycles have not
been undertaken because an advantage cannot be seen. The data acquisition is not
easier from space than from ground, the instruments are fairly complicated and need
maintenance.

Measurements of the solar irradiance are different. These are truly space experiments
because the changes of irradiance are subtle (< 1 percent) and, most important, the
Earth's atmosphere limits substantially the capability of measuring the total solar
output to the required accuracy (Fröhlich, 1977, Willson, 1985). There is substantial
scientific impetus to monitor continuously the effects of solar internal dynamics
by means of a radiometer. As I understand it, SOHO will give the opportunity for
this purpose for an extended period of time.

Helioseismology is a rapidly growing field of research. Observations from the ground
have already demonstrated the great potential of probing the solar interior with
the aid of global oscillations (see e.g. Gough, 1985, and Noyes and Rhodes, 1984,
for reviews). The frequencies have to be measured as accurately as possible with this
technique, but several obstacles hamper its full exploitation. Firstly, the data

string should cover a time span as long as possible on a duty cycle as complete as possible. Night gaps must be avoided. Secondly, stochastic motions on the Sun produce noise which can be reduced again by long integration. And thirdly, seeing effects deteriorate the signal of the high degree (1 $\gtrsim$ 300) modes, as reported by Noyes and Rhodes (1984).

There is then obviously a strong desire to have an observational facility in space with full sunlight for several years. The SOHO spacecraft at the inner Lagrange point will offer this opportunity. A non-imaging velocity-meter is on its payload list with highest priority. This will be an instrument with a rather low data rate. (I have learned during this conference that Earth-based networks with non-imaging resonance cells are under construction by a french group in Nice and in a collaboration of scientists from Birmingham, U.K., and from the Canary Islands, Spain.) Imaging instruments are desirable to analyse the medium and high degree modes. Their data production is certainly much higher than that of non-imaging cells. Noyes and Rhodes speak of a 1024 x 1924 array for the full disc and of 10^{10} samples per day for a space mission.

As far as ground based observations are concerned, there is the project of the Global Oscillations Network Groups (GONG), housed at Kitt Peak and with much support from all over the world. To have a sunshine coverage as complete as possible 6 stations distributed over the globe are planned with 256 x 256 pixel images of the full disc. There one speaks of a data flow of 200 M bytes per day and station.

5. CONCLUSIONS

The High Resolution Solar Observatory (HRSO) on board of Space Shuttle will be the next space facility for high spatial resolution. The scientific return is expected to be very high. Its flight duration will be limited to about 10 days. The capability of HRSO will be increased very much if it is put on a full sunlit orbit. Moreover, it should preferably fly during the years when SOHO is operating its high resolution UV and XUV spectrometers - although delays of one or the other spacecraft flights on technical and financial grounds make the coordination very difficult. Yet in this way HRSO will provide a link between the seismological studies of the inerior of the Sun and the coronal and wind investigations by SOHO.

The Large Earth-bound Solar Telescope will be much more versatile than HRSO. There exists a very broad variety of targets for LEST during its long life-time. Its post-focal instrumentation may be changed and improved continuously. We have to await further development until we know to which degree a telescope like LEST can be made free from seeing.

What about the era beyond LEST, HRSO, and GONG? How much solar observation will be possible on COLUMBUS/Space Station/EURECA? Will it be possible, starting with small steps and going to a more distant future in several decades, to shift much of the activities from ground to space and to have there an universal instrumentation with an observatory status which allows repair, servicing, replacement of components, and the acquisition of large amounts of data? The advantages of such a facility are clear. We should keep us informed about the possibilities to come and representatives from ESA might encourage us to work for its realisation.

Acknowledgements: I am grateful to Dr. A. Wittmann for helpful discussions.

REFERENCES

Ando, H., Osaki, Y.: 1977, Publ. Astron. Soc. Japan 29, 221
Antia, H.M., Chitre, S.M., Narasimha, D.: 1984, in "Solar Seismology from Space", R.K. Ulrich, J. Harvey, E.J. Rhodes, Jr., J. Toomre, eds., JPL Publication 84-84, Pasadena, p. 345
Blackwell, D.E., Dewhirst, D.W., Dollfus, A.: 1959, Monthly Notices Roy. Astron. Soc. 119, 98
Bogdan, T.J., Zweibel, W.G.: 1985, Astrophys. J. 298, 867
Danielson, R.E.: 1961, Astrophys. J. 134, 275
Deinzer, W., Knölker, M., Voigt, H.H. eds.: 1986, proceedings of a workshop on "Small Scale Magnetic Flux Concentrations in the Solar Photosphere", Vandenhoeck and Ruprecht, Göttingen
Deubner, F.-L.: 1976, Astron. Astrophys. 51, 189
Durrant, C.J., Nesis, A.: 1982, Astron. Astrophys. 111, 272
Fröhlich, C.: 1977, in "The Solar Output and Its Variation", O.R. White, ed., Col. Ass. Un. Press Boulder, p. 93
Gingerich, O.J., Noyes, R.W., Kalkofen, W., Cuny, Y.: 1971, Solar Phys. 18, 347
Goldreich, P., Keeley, D.A.: 1977, Astrophys. J. 212, 243
Gough, D: 1980, in "Nonradial and Nonlinear Stellar Pulsation", H.A. Hill and W.A. Dziembowski, eds., Springer, Berlin, p. 273
Gough, D.: 1985, Solar Phys. 100, 65
Hasan, S.S.: 1984, Astrophys. J. 285, 851
Heath, D.F., Thekaekara, M.P.: 1977, in "The Solar Output and Its Variation". O.R. White ed., Col. Ass. Un. Press Boulder, p. 193
Howard, R.: 1985, Solar Phys. 100, 171
Keil, S.L. ed.: 1984, proceedings of NSO conference on "Small-Scale Dynamical Processes in Quiet Stellar Atmospheres", Sunspot, N.M.
Kneer, F., Wiehr, E.: 1987, in preparation
Krat, V.A.: 1981, Solar Phys. 73, 405
Krat, V.A., Karpinsky, V.N., Sobolew, V.M., Dulkin, L.Z., Motenko, B.N. Khalezov, P.A.: 1970, Izv. Pulkovo 185, 124
Lamla, E.: 1982, in Landolt-Börnstein, "Numerical Data and Functional Relationships in Science and Technology", Vol. VI /2b
Landi degl'Innocenti, E.: 1985, in "Theoretical Problems in High Resolution Solar Physics", H.U. Schmidt ed., MPA 212, München, p. 162
Mehltretter, J.P.: 1978, Astron. Astrophys. 62, 311
Mein, N., Schmieder, B.: 1981, Astron. Astrophys. 97, 310
Muller, R.: 1985, Solar Phys. 100, 237
Nesis, A.: 1986, IAU Symp. 123, Aarhus
Noyes, R.W., Rhodes, E.J., Jr.: 1984, "Probing the Depths of a Star: the Study of

Solar Oscillations from Space", JPL, Pasadena, Cal.
Pahlke, K.D., Solanki, S.K.: 1986, Mitt. Astron. Ges. 65, 162
Parker,, E.N.: 1978, Astrophys. J. 221, 368
Pierce, A.K., Allen. R.G.: 1977, in "The Solar Output and Its Variation", O.R.
    White ed., Col. Ass. Un. Press Boulder, p. 139
Plass, G.N., Yates, H.: 1965, Handbook of Military Infrared Technology, W.L. Wolfe
    ed., Nav.Res.Dep. of the Navy, Washington, D.C., p. 175
Roudier, Th., Muller, R.: 1986, Solar Phys. 107, 11
Scharmer, G.: 1987, in "The role of fine-scale magnetic fields on the structure of
    the solar atmosphere", proceedings of inaugural workshop, Tenerife, 1986
Scholiers, W., Wiehr, E.: 1985, Solar Phys. 99, 349
Schröter, E.H.: 1985, Solar Phys. 100, 141
Skumanich, A., Smythe, C., Frazier, E.N.: 1975, Astrophys. J. 200, 747
Staiger, J.: 1987, Astron. Astrophys. 175, 263
Stenflo, J.O. ed.: 1983, IAU Symposium No. 102, "Solar and Stellar Magnetic Fields:
    Origins and Coronal Effects", Reidel, Dordrecht
Stenflo, J.O., Harvey, J.W., Brault, J.W., Solanki, S.: 1984, Astron. Astrophys.
    131, 333
Title, A.: 1985, in "Theoretical Problems in High Resolution Solar Physics", H.U.
    Schmidt ed., MPA 212, München, p. 28
Vernazza, J.E., Avrett, E.H., Loeser, R.: 1976, Astrophys. J. Suppl. 30, 1
von Uexküll, M., Kneer, F., Mattig, W., Nesis, A., Schmidt, W.: 1985, Astron. Astro-
    phys. 146, 192
von der Lühe, O.: 1987, in "The role of fine-scale magnetic fields on the structure
    of the solar atmosphere", proceedings of inaugural workshop, Tenerife, 1986
Wiehr, E.: 1987, private communication
Willson, R.C.: 1985, quoted by V. Domingo and J. Ellwood in "Solar and Heliospheric
    Observatory", ESA SCI(85) 7
Wittmann, A.D., Xu, Z.T.: 1987a, Astron. Astrophys. Suppl., in press
Wittmann, A.D., Xu, Z.T.: 1987b, Proc. NATO Advanced Res. Workshop "Secular Solar
    and Geomagnetic Variation", in press

NEW IDEAS ABOUT GRANULATION BASED ON DATA FROM THE SOLAR OPTICAL UNIVERSAL POLARIMETER INSTRUMENT ON SPACELAB 2 AND MAGNETIC DATA FROM BIG BEAR SOLAR OBSERVATORY

A.M. Title, T.D. Tarbell, K.P. Topka, R.A. Shine
Lockheed Research Laboratory
Palo Alto, California USA

G.W. Simon
Air Force Geophysics Laboratory
Sunspot, New Mexico USA

H. Zirin
California Institute of Technology
Pasadena, California USA

and the SOUP Team

ABSTRACT

The Solar Optical Universal Polarimeter (SOUP) on Spacelab 2 collected time sequences of diffraction limited (0.5 arc second) granulation images with excellent pointing (.003 arc seconds) and freedom from the distortion that plagues groundbased images. The solar 5 minute oscillations are clearly seen in the data. Using Fourier transforms in the temporal and spatial domains, we have shown that oscillations have an important effect on the autocorrelation (AC) lifetime. When the oscillations are removed the autocorrelation lifetime is found to increase from 270 seconds to 410 and 890 seconds in quiet and magnetic regions, respectively. Exploding granules are common and it is hard to find a granule that neither explodes nor is unaffected by an nearby explosion. We speculate that a significant fraction of granule lifetimes are terminated by nearby explosions. Via local correlation tracking techniques we have been able to measure horizontal displacements, and thus transverse velocities, in the intensity field. It is possible to detect both super and mesogranulation. Horizontal velocities are as great as 1000 m/s in quiet sun and the average velocity is 400 m/s and 100 m/s in quiet and magnetic sun, respectively. These flow fields affect the measured AC lifetimes. After correcting for steady flow, we estimate a lower limit to the lifetime in quiet and magnetic sun to be 440 and 950 seconds, respectively. The SOUP flow fields have been compared with carefully aligned magnetograms taken at the Big Bear Solar Observatory (BBSO) before, during, and after the SOUP images. The magnetic field is observed to exist in locations where either the flow is convergent or on the boundaries of the outflow from a flow cell center. Streamlines calculated from the flow field agree very well with the observed motions of the magnetic field in the BBSO magnetogram movies.

INTRODUCTION

Until the flight of the Solar Optical Universal Polarimeter(SOUP) on Spacelab 2, solar granulation was considered to be convective overshoot in the outer layers of the sun, in which the convective cells had a center-to-center distance of about 2.5 arc seconds and an autocorrelation (AC) lifetime of 3 to 7 minutes. Lifetimes from tracking of individual features varied from a few to 45 minutes. There had been some indication that granulation was slightly different in magnetic field regions and near sunspots, but there was controversy on this point. On average the picture was that a granule was a bright (hotter than average) region of upflow, with a diameter of 1.5 arc seconds, surrounded by a darker (cooler) region of downflow. The status of granulation before the flight of SOUP was excellently reviewed in "Solar Granulation" by Bray, Loughhead, and Durrant (1984).

The picture of granulation emerging from the analysis of the SOUP data is qualitatively different from the traditional model. We still cannot fully characterize granule evolution, but we can describe the phenomena of exploding granules and some associated effects. A complete evolutionary description is complex because there exists in the solar surface a hierarchy of intensity fluctuations, from the f and p–mode oscillations (with scales of a few to many tens of arc seconds and large phase velocities) to local phenomena (which involve areas with radii of a few to ten arc seconds). Exploding granules, which in previous data were relatively rare, are seen to be pervasive in SOUP data. All of the wave fields and the local phenomena, of course, coexist in the intensity field we have called granulation, so that a single image does not begin to reveal the richness in the solar surface.

In addition the solar photosphere exhibits systematic flows with spatial scales from 10 to at least 40 arc seconds, and lifetimes long compared to the thirty minute observing period imposed by the Spacelab 2 orbit. Some of the flow fields are the meso and supergranulation, but other patterns may also exist. The existence of the flow fields can have a significant effect on the measured lifetime of the granulation pattern. Our results on pattern lifetime are preliminary, but clearly indicate that previous estimates by correlation techniques have underestimated the lifetimes of the individual features. Our measurements in magnetic areas also show significant differences from quiet regions.

Using doppler imaging techniques, Simon and Leighton (1964) first observed the solar supergranulation flow field. They showed that the supergranulation flow was generally inside the solar network pattern observed in Ca II K and a number of other temperature sensitive solar lines. Leighton (1964) later conjectured that the flow field carried the solar magnetic field to the flow cell boundaries. He also created a diffusion model of the solar cycle in which the interaction of the flow field and the magnetic field was responsible for the distribution of the field along the solar surface. There have been a number of attempts to measure the motions of individual field elements (Smithson (1973), Mosher (1977), and Martin(1987)) in order to validate the concept that the field was moved to the supergranulation boundaries. The recent observations of the motion of intranetwork fields at BBSO (Martin, 1987) have shown motion of mixed polarity area toward the flow boundaries.

Because doppler imaging is only sensitive to the line of sight velocity and the flow pattern in the supergranulation is largely parallel to the solar surface, attempts to detect the vertical component of the supergranulation at disk center have met with only marginal success (Simon and Worden(1976)). The granulation pictures taken in space by the SOUP instrument on Spacelab 2 allowed a different approach to detecting surface flow perpendicular to the line of sight - direct displacement measurements of the granulation pattern.

DATA

The granulation data were collected on film using the Solar Optical Universal Polarimeter (SOUP) which operated on the flight of Spacelab 2 (Title, *et al.* 1986). The original images are 140 x 250 arc seconds and are taken in a thousand angstrom band centered at 6000 Å. The images are very uniform in quality and are distortion-free. Using the Spacelab Instrument Pointing System (IPS) and an internal fast guider, image stability of 0.003 arc second RMS was achieved. For this preliminary report most of the results are based on studies of three digitized subsections, a 40 x 40 arc second quiet sun region, a 60 x 60 arc second pore region, and a 100 x 100 arc second region centered on AR 4682 at approximately S15, W31 ($\mu = 0.75$). The pores are part of AR 4682. Because of limitations of the processing computer and our image display system, all digital images are 256 x 256 pixels. This corresponds to 0.161, 0.231, 0.381 arc second per pixel for the quiet, pore and active regions, respectively. The images are separated by 10 seconds in time and cover 1650 seconds. The results reported here are all from orbit 110 (19:10:35 to 19:38:05 GMT, on 5 August 1985.)

During the flight of Spacelab 2 Big Bear Solar Observatory (BBSO) collected correlative data on AR 4682. The BBSO data included magnetograms, H_α, and Ca II K images. The BBSO data covered the period before, during, and after the SOUP maps. For this report only the medium sensitivity (1024 integrations) magnetograms have been used.

RESULTS

Temporal Autocorrelation Functions

One of the standard methods for determining the lifetime of an intensity pattern is from the width of its temporal autocorrelation (AC) function. Here we define the temporal autocorrelation of the intensity as

$$AC(\tau) = \frac{\langle \delta I(x,y,t_o)\, \delta I(x,y,t_o + \tau)\rangle}{\langle \delta I^2(x,y,t_o)\rangle}$$

where x and y are the spatial coordinates, t_o is the reference time, and τ the time separation. The brackets indicate an average over space, and

$$\delta I(x,y,t) = I(x,y,t) - \langle I(x,y,t)\rangle$$

where here the brackets indicate an average over space and time.

Previous AC lifetime measurements of granulation have ranged from 3 to 7 minutes with an average of about 6 minutes. The lifetime, as measured by the time for the correlation to drop to $1/e$ (0.37) is, from figure 1a, about 5 minutes.

Our data for figure 1a were obtained from an area of about 300 square arc seconds . When smaller areas are examined the AC functions look quite different. Figure 2 (solid) shows the AC functions generated from four 36 square arc second areas. In these smaller regions the AC functions do not drop monotonically, but rather exhibit oscillations with periods of 3 to 6 minutes (depending on the region). This suggests that the five minute oscillations are affecting the correlation lifetime, which is not too surprising as it is strongly present in the granulation movies. The average AC over a sufficiently large region does not oscillate as strongly as a small region because a range of periods and modes are present.

To remove the effects of the five minute oscillation we have applied what we call a subsonic Fourier filter to the time sequence of quiet sun images. The original sequence of images is Fourier transformed from a function of x, y, and t into a transformed function of k_x, k_y, and ω. The subsonic filter is defined by a cone

$$\omega = v \times k$$

in $k - \omega$ space, where k and ω are spatial and temporal angular frequencies and v is a velocity. All fourier components inside the cone (i.e., with phase velocities less than v) are retained, while all those outside are set to zero. Then a new sequence of images is calculated by an inverse Fourier

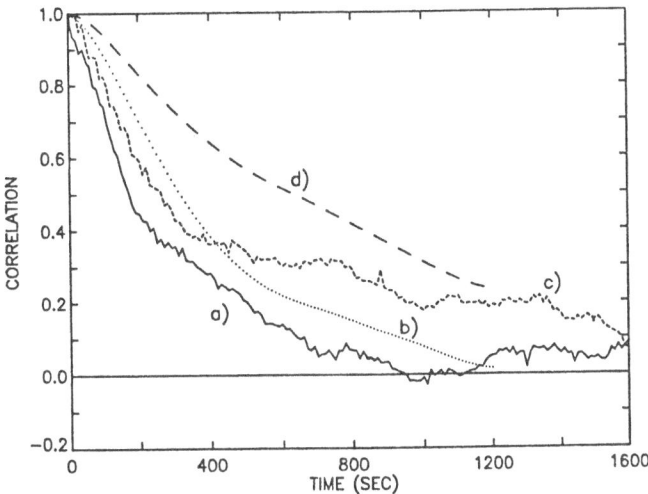

Fig. 1 Autocorrelation measurements from quiet sun original (a), Fourier filtered data (b), magnetic region original (c), and Fourier filtered data (d).

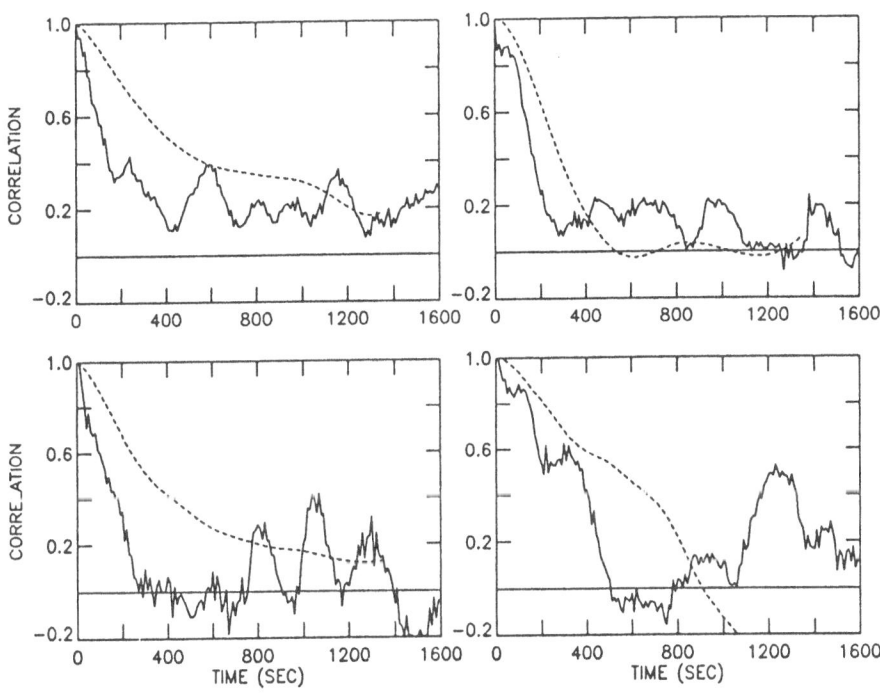

Fig. 2 AC(τ) in four 6 × 6 arc second regions (solid). These exhibit oscillations due to the solar p-modes. Dashed lines show AC(τ) for the same regions after Fourier filtering.

transform. A value of $v = 3$ km/s, well below the sound speed of 7 km/s, is used. For this value of v, the velocity cone is totally inside of the the five minute oscillation modal distribution.

A subsonic filtered movie exhibits very little five minute oscillation. As seen from figure 2 (dashed) the AC functions created from subsonic filtered data for the individual regions have the oscillations

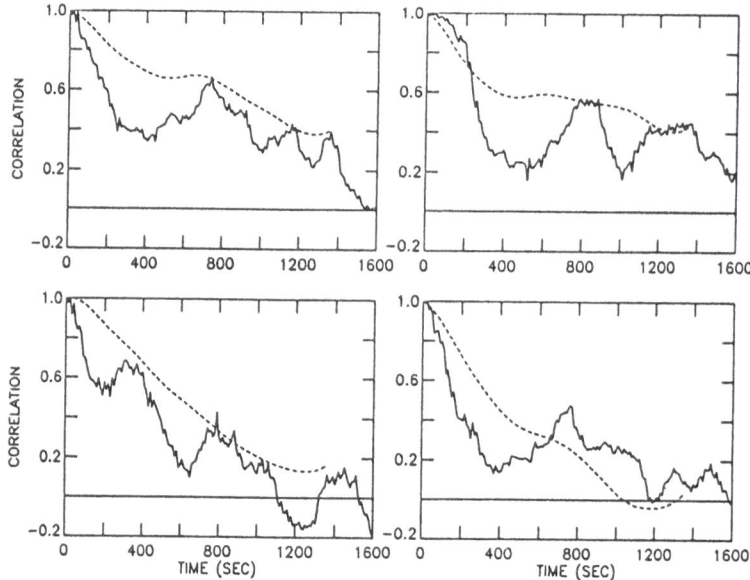

Fig. 3 AC functions for four magnetic regions using original (solid) and Fourier filtered data (dashed).

strongly suppressed. Figure 1 shows the AC of 300 SAS for original (a) and subsonic filtered (b) images. The AC lifetime of the subsonic data is 1.5 times greater than the original – a lifetime of 410 versus 270 seconds.

In order to observe the effect of magnetic field on the granulation pattern the SOUP images have been aligned with National Solar Observatory (Tucson) magnetograms taken just before and just after the SOUP data. Twelve 36 square arc second regions were selected to be inside the 70 gauss contour of the magnetograms, but well outside of pores. Shown in figure 3 are the AC functions for the original (solid) and the subsonic filtered (dashed) data for four of these regions. Figure 1 shows the sum of the AC's for the twelve regions for original (c) and subsonic filtered (d) data in the magnetic areas. The lifetime from the subsonic data is 2.1 times the original data – 890 versus 420 seconds.

Spatial Autocorrelation Functions

The temporal autocorrelation functions discussed above compare sets of spatially aligned data sets as a function of time and yield information related to feature lifetime. Spatial autocorrelation functions compare the same image with different spatial offsets and yield information on the spatial scale of the image. Here we define the spatial AC function as

$$AC(\Delta) = \frac{\langle \delta I(x,y,t) \; \delta I(x+\Delta,y,t) \rangle}{\langle \delta I^2(x,y,t) \rangle}$$

where Δ is the spatial offset.

For a wide variety of patterns the half width at half maximum (HWHM) of the correlation function is related to the pattern size, r_0, by

$$r_0 = 1.2 \times \text{HWHM}.$$

From radius of the average structure in the quiet and magnetic regions is 450 and 500 km (diameters of 1.25 and 1.4 arc seconds), respectively, which is consistent with previous measurements. The structures in the quiet and magnetic regions are about the same size.

The spatial AC functions can also be used to understand the effect of flows on the lifetime data because the displacement field can be interpreted as the result of a flow field. That is,

$$\Delta = v \times t.$$

The flow lifetime, T_f, the time required for the flow to cause the spatial AC to drop to $(1/e)$ is just

$$T_f = \frac{AC_{1/e}}{v_f},$$

where v_f is the flow velocity. The value of $AC_{1/e}$ is 450 km for 500 km for quiet sun and magnetic sun, respectively.

The presence of a flow field will cause the measured lifetime from the temporal AC function to underestimate the lifetime of the granulation. Roughly, the lifetimes as measured by the temporal AC functions, the actual lifetime, and the flow lifetime from the spatial AC functions are related by

$$1/T_m = \sqrt{(1/T_a)^2 + (1/T_f)^2},$$

or

$$T_a = \frac{T_m \times T_f}{\sqrt{T_f^2 - T_m^2}},$$

where T_a is the actual lifetime of the granules and T_m is the measured lifetime from the temporal AC function. Since the flow lifetime must be greater than the measured lifetime, we can set an upper limit to the average flow velocity. For quiet and magnetic sun the maximum flow velocities permitted by the observations are 1.1 and 0.5 km/s, respectively.

Via local correlation techniques (November, et al. 1986), we have measured the average flow velocities for quiet and magnetic to be 400 and 200 m/s, respectively. Correcting for the flow field

Table 1.

	Original	Subsonic	Steady Flow Correlation	Feature Flow Correlation
Quiet	270s	410s	440s	710s
Magnetic	420s	890s	950s	–

yields "corrected" quiet and magnetic granule lifetimes of 440 and 950 seconds, respectively. The AC lifetime data is summarized in Table 1.

Granule Evolution

Once the oscillations are removed, it is obvious from the movies that exploding granules are a major component of the evolutionary history of the local and overall flow. A significant fraction of granules are seen to explode, and those that don't explode have their evolutionary history dominated by exploders in their near vicinity. Waves generated by exploding granules can clearly be seen to travel at least the 2.5 arc second separation distance of the granulation pattern, and there is some evidence that they may propagate much further.

Because of the confusion caused by the multiple interactions of exploding granules and their wave fields, it is difficult to develop statistical measures that are characteristic of a single pattern. Rather the statistics reflect that on some spatial and temporal scales the flow is convective and on others it is turbulent. It is possible, however, to infer a picture of the evolution of a prototypical exploding granule, based upon the appearance in movies and guided by model calculations (Nordlund, 1983) of granule behavior. In this conception an exploding granule is caused by a hot plume which rises because it is less dense than the surrounding gas. When the plume reaches the surface it radiates its excess thermal energy and cools. The cooler gas contracts and loses its buoyancy. It then falls onto the rising plume below it. The gas below is deflected horizontally, causing a hot annulus to form. Finally, the horizontally expanding front generates surface waves.

When seen from above, initially the hot plume appears to be a bright, roughly circular region on the surface. As the spot cools, the center of circle dims causing an annular region to appear. However, as the annulus expands, its width decreases and the thin expanding region is broken apart by the local surface pressure fluctuations. Moving ahead of the annulus is a surface wavepattern that causes the appearance of rows of thin, elongated granules to appear.

In a typical event, the bright ring covers the locations of the nearest and infringes upon the next-nearest neighbor granules. The nature of the interaction of the expanding bright front and the local surroundings is difficult to describe under all circumstances. Because exploding granules are so common, a significant fraction of the surface area is dominated by these horizontally flowing fronts. The neighboring granules then must also be deflected or must deliver their excess thermal energy somewhat below the surface. The expanding fronts move at speeds of 0.5 to 1.5 km/s and therefore take between 500 to 1500 seconds to expand an arc second in radius. The surface waves are much faster, about 7 to 10 km/s.

The movies last only 1600 seconds so that it is difficult to say what occurs after an explosion, as an event and the length of the movie are comparable. Exploding granules are only recognized clearly if the original source can be recognized.

Although it is possible to describe an exploding granule event fairly well, there are other phenomena occurring whose history is much harder to decode. Multiple exploding granules are the least confusing of the complex events. What appears to occur is that a group of granules explodes as one, creating an expanding front that is an irregular oval. The oval events are often recognized after they have started, that is, evolution from a set of bright sources is not always seen.

Relations Between Flow and Magnetic Fields

Shown in Figure 4 are (a) a SOUP image, (b) a BBSO magnetogram at nearly the same time, (c) the SOUP flow field (shown as vectors) overlayed on a gray-scale map of the divergence of the flow field, and (d) the SOUP flow field superposed on the magnetogram. The speed and also the direction of the transverse motions are accurately measured by the correlation tracking algorithm. Values generally lie in the range 100 to 800 m/s, with one-sigma uncertainties of about 75 m/s. The velocities (shown as arrows in Figure 4) represent the average velocity obtained by correlation tracking over the 28 minute observation time of orbit 110. Since the divergence of the horizontal flow vector u is associated with upflows and downflows through the mass continuity equation (November, *et al.* (1986)), the divergence map accurately portrays the average vertical velocity. Thus, it clearly identifies cell interiors (sources or upflows) and boundaries (sinks or downdrafts).

We see from the figure that the flow field and the magnetic field are intimately related. In regions where the magnetic field is in roughly linear structures the flow is directed toward these structures. In relatively compact areas the flow field points toward the concentrations. In cell-like regions of the magnetogram the vectors of the flow field point radially outward from the cell centers toward the boundaries (network).

Additional insight into the relationship of flows and magnetic fields is gained by asking where the flow field would carry free particles ("corks") that are originally distributed uniformly in the

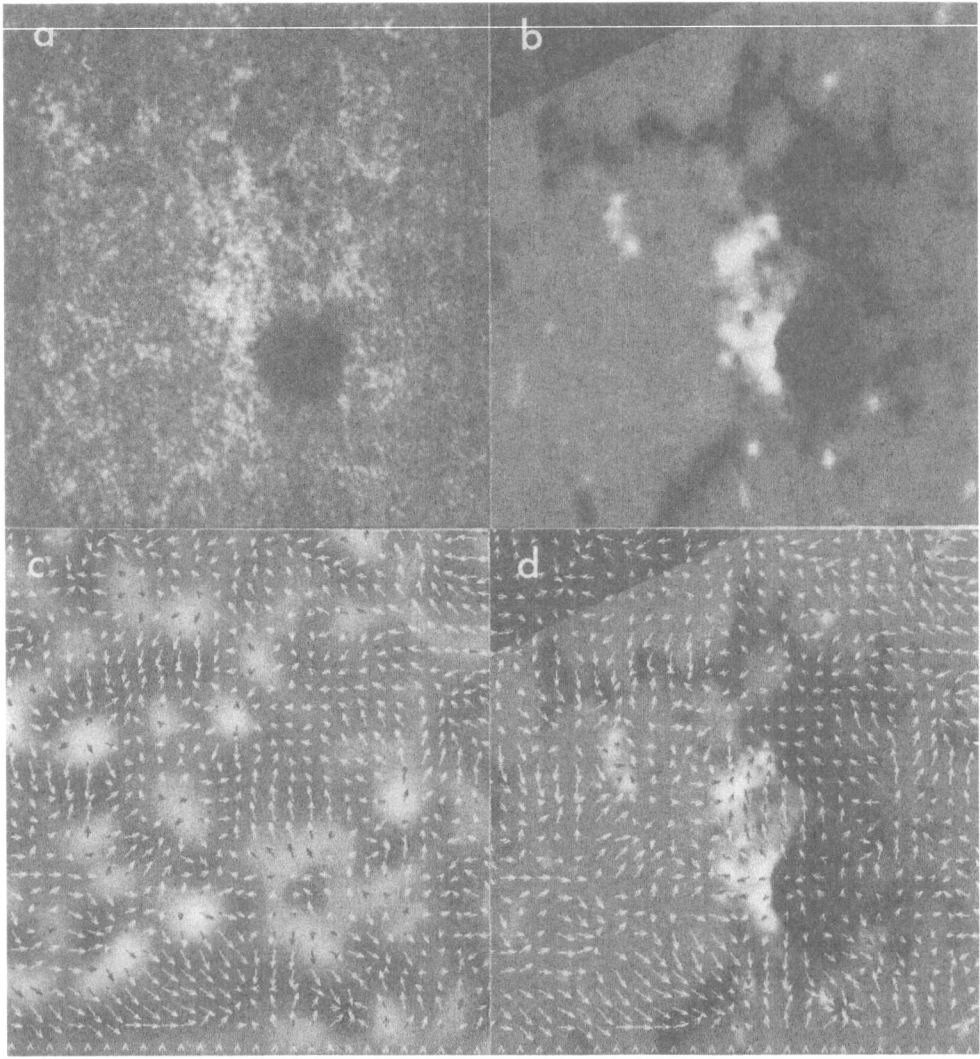

Fig. 4 The relation of flow and magnetic fields; (a) a SOUP image, (b) a BBSO magnetogram at nearly the same time, (c) the SOUP flow field overlayed on the divergence of the flow field, and (d) the flow field overlayed on the magnetogram.

flow field. The cork flow is calculated by moving the corks according to the velocity of the local flow field. Figure 5 shows (a) an initial uniform distribution of corks on a SOUP image, (b) the corks overlayed on a magnetogram after 4 hours, (c) 8 hours, and (d) 12 hours. The cork motion calculation assumes that the flow field remains unchanged over a 12 hour period. The 75 m/s uncertainty in the flow speeds used for Figures 4 and 5 would cause an uncertainty of less than 5 arcsec in cork positions after 12 hours. This is quite small compared to the overall scale of the final cork pattern. As a check the flow fields were also calculated from shorter (and thus noisier) data

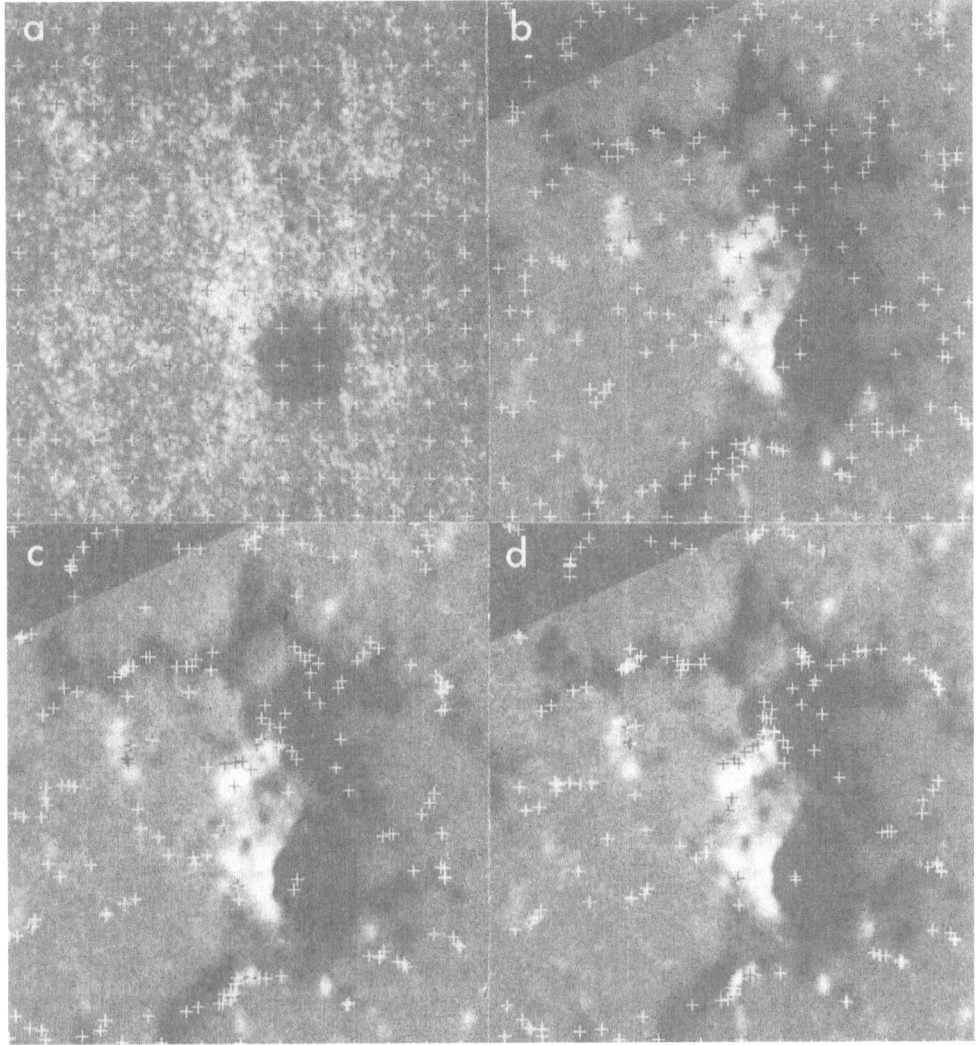

Fig. 5   The evolution of a uniform distribution of free particles "corks"; (a) the initial uniform distribution of corks on a SOUP image, (b) a magnetogram overlayed with the corks after 4 hours, (c) 8 hours, and (d) 12 hours.

sets obtained in orbits 108, 109, and 111, which span an interval of 4.5 hours. These measurements do not show convincing evidence for major changes in the average flow field over this span, but unfortunately they are severely limited by "solar noise" contributed by random motions of granules and by five-minute oscillations.

The cork paths represent streamlines of the flow. The streamlines and corks illustrate an important feature of the flow pattern. Initially the flow carries corks to the cell boundaries, but then, as time proceeds, the corks are carried along the boundaries to sink regions. These sinks are usually vertices in the network pattern as seen in calcium, hydrogen, or magnetograms. In the course of the data analysis a magnetogram movie was made which was overlayed with the SOUP surface flow field vectors. It was very clear from the movie that magnetic field elements usually move first to cell boundaries and then flow along the boundaries just as the cork model suggests.

Both the magnetograms and the cork simulation show an incomplete network: fully-outlined cells are rare, and usually there are just enough markers in the boundaries to suggest a cellular pattern. The cork simulation shows that this is an intrinsic property of the flow patterns and not simply a result of insufficient magnetic flux to complete the pattern. The magnetic (or cork) network also suggests a larger dominant cell size than do the flow maps. Cell sizes (center-to-center) in the flow divergence maps are typically 10,000 to 15,000 km rather than the 30,000 km value usually associated with supergranulation.

CONCLUSIONS

This is a preliminary report. We have corrected AC lifetimes for the 5 minute oscillation and large scale flows, but we have neglected more local flows caused by exploding granules and other phenomena which we know exist. Nonetheless, these corrections have increased the statistically measured lifetime by a factor of 1.6 and 2.25 in quiet and magnetic sun, respectively. In magnetic regions the uncorrected AC lifetime is dominated by the five minute oscillation. When this oscillation is removed the AC technique demonstrates that the average intensity pattern is significantly more stable in magnetic field regions. Although the surface intensity patterns are considerably less vigorous in magnetic compared to non-magnetic regions, visually they are nearly as complex. The granulation pattern is statistically more stable, there is lower modulation in the intensity field, the wave patterns are diminished in amplitude as are the velocities of the steady flows, and exploding granules are much less common; but most of the phenomena seen in quiet sun can also be found in the magnetic regions.

We believe that the surface flow as indicated by the local intensity pattern is closely related to the motion and evolution of the magnetic field. This implies that the granulation pattern is advected by mesogranular and supergranular flow fields. It is apparent that the flow field determined from one 28 minute measurement is an excellent surrogate for the topology and motion of the corresponding magnetic field configuration, and is valid for at least four hours prior to and after the SOUP observations. From the cork simulations we estimate that the magnetic pattern would require about 8 to 10 hours to develop. This suggests that the flow field and magnetic field have a lifetime considerably longer than 12 hours as would be expected for large-scale supergranular structures. Once the field gets to a boundary the flow velocity slows considerably.

Our observations indicate that flow along network boundaries may be an important feature in the evolution of the magnetic field pattern. This would have important implications for coronal heating and buildup of magnetic stresses in the network. First, flow along the network boundaries will tend to mix and twist the fields on very small scales. This mixing will be enhanced by local displacements of the field caused by randomly-directed motions and explosions of individual granules. Since both dissipation and heating in magnetic regions depend critically on the spatial scale of the twisting of the flux tubes, chromospheric and coronal heating can be enhanced by the flow along the boundaries. Second, flow along the boundaries concentrates fields in vertices. These vertices are probably stable points in the flow field, so that new supergranules may form with a vertex at the previous boundary. If so, the random-walk diffusion of magnetic field discussed by Leighton (1964) may be much lower than would otherwise be expected.

Figures 4 and 5 are in excellent agreement with Simon and Leighton's (1964) idea that the flow field pushes the magnetic field into boundaries. However, it is observationally difficult to distinguish this concept from the converse hypothesis that the locus of the flow field is constrained by the prior presence of magnetic field (the traditional chicken-or-the-egg conundrum!).

Special thanks are extended to the crew of Spacelab 2 and the controllers and planners on the ground who worked so hard to get the observations described here on the extension day of the mission. The SOUP instrument was built by and under the direction of Mike Finch, Gary Kelly, Roger Rehse, and Ralph Reeves. The SOUP experiment was supported by NASA under contract NAS8-23805. The image processing developments using laser optical disks have been supported by Lockheed Independent Research funds.

REFERENCES

Bray, R.J., Loughhead, R.E., and Durrant, C.J, 1984: *"The Solar Granulation,"* Cambridge University Press.

Leighton, R., 1964: *Ap. J.*, **140**, 1559.

Martin, S., 1987: *Solar Physics,* in press.

Mehltretter, J.P., 1978: *Astr. Astrophy.*, **62**, 311.

Mosher, J.M., 1977: PhD. Thesis, California Institute of Technology.

Namba, O., 1986: *Astr. Astrophy.*, **161**, 31.

Nordlund, A. 1983, *IAU Sym. 102*, 79.

November, L.J., Simon, G.W., Tarbell, T.D., Title, A.M., and Ferguson, S.H.: 1986, "Large-Scale Horizontal Flows from SOUP Observations of Solar Granulation," *Proceedings of the* Second Workshop on Problems in High Resolution Solar Observations, Boulder, CO, September 1986.

Simon, G.W., and Leighton, 1964: *Ap. J.*, **140**, 1120.

Smithson, R., 1973: *Solar Physics*, **29**, 365.

Title, A.M., Tarbell, T.D., Simon, G.W., and the SOUP Team, 1986: *Advances in Space Res.*, in press.

Wittmann, A., 1979: *"Small Scale Motions on the Sun,"* Kiepenheuer-Institut für Sonnenphysik.

Worden, S.P., and Simon, G.W., 1976: *Solar Physics*, **46**, 73.

FUTURE PROSPECTS OF STELLAR AND SOLAR PHYSICS FROM SPACE

R.-M. Bonnet
European Space Agency
8-10 rue Mario Nikis, 75738 Paris Cedex 15, France

Abstract

The unbeatable advantage of Space techniques in the investigation of solar and stellar physics are briefly reviewed, and a Table listing most of the missions and facilities which are foreseen to contribute to the field in the next 15 years or so is presented. We then review different scientific areas for which we think Space Techniques will provide substantial advance. This is astrometry, the investigation of stellar and solar interiors, of external atmospheres and winds, of solar and stellar flares, the observations and studies of stars at various phases of their evolution and the search for very high resolution imagery.

Some specific missions are described, providing a concrete illustration of the extreme wealthiness and of the potential of this exciting field of Science.

I   Introduction

The utilisation of Space techniques in Astrophysics and in particular, in the area of stellar and solar physics has proved to offer unbeatable advantages and has led to considerable progress which makes them today indispensible in the perspective of the evolution of this domain of science.

The elimination of atmospheric absorption allows the observation of the entirety of the electromagnetic spectrum from gamma-rays to the far infrared and the radio domain. At the same time, the elimination of all other atmospheric perturbations like proper light emission and turbulence allows absolute radiometry and provides images whose quality should only in principle be limited by the quality of telescopes, thereby allowing the sharpest view of the Universe.

Since there is no limitation on where to place a telescope in Space, it is in principle possible to optimize the observing site to the observation or programme of a given instrument. Access to new and exceptional observing sites is therefore permitted, such as the L_1 and L_5 Lagrangian points of the Sun-Earth-Moon system which offers 24 hours of either full sunlight or complete darkness observations. Special sites like the geostationary orbit of the International Ultraviolet Explorer (IUE) permit simultaneous utilisation by astronomers of two separated continents. The high excentricity of the EXOSAT orbit was the reason for many discoveries of neutron star emissions and quasi periodic objects, due to the long and uninterrupted observing time. The Moon itself may offer in the future, unique observing sites in particular for infrared and radio astronomy.

Space techniques permit in situ observations of the interplanetary medium, the planets and their atmosphere, but also of their soil, of comets and even of the Sun itself. The Pioneer 11 Satellite will very soon reach the limits of the heliosphere and continue its journey in interstellar space.

Finally, in the not so distant future, we may see the deployment of ultra long baselines for Interferometry, first in the radio domain, where it is presently the Earth's diameter which imposes limitations to the use of this technique, and later on in the optical and ultraviolet regimes.

From the preceding, one can guess that major progress is expected in the near or longer term. However, there is one drawback to the use of space techniques which has been dramatically illustrated by the recent accidents of the Challenger Space shuttle and of the Ariane European launcher, the consequences of which have in the best case, grounded if not cancelled definitely major astronomical missions such as the US-European Hubble Space Telescope (HST), the German-US ROSAT, the European Hipparcos, without forgetting the sadly affected Ulysses mission. Consequently, it is very difficult if not impossible, to forecast today the precise succession of future astronomy missions. In the remainder of this paper, we will describe some of these missions. Any mention of their possible launch date or launch means should be considered as tentative and likely to be subject to future changes.

II   Review of the major future astronomy facilities

We present here in synoptic form the list and main characteristics of solar and stellar missions which have already been developed or are in the process of being developed.

Table 1 gives as an exhaustive list as possible of all missions of interest for stellar and solar physics which are already under development or envisaged in the timeframe between now and 2000, in Europe, Japan, the United-States, and the USSR. The location of the missions in the Table, depends on both the part of the electromagnetic spectrum for which the mission is designed and the launch date as known today. Interrogative marks are indicative of either the uncertainty of this date, or of the status of programmation of the mission itself. Acronyms are explained in the text of the subsequent chapters. Figures 1 and 2 compare the relative sensivities of some of the instruments listed in Table 1 and Figure 3 compares the angular resolution for some of them.

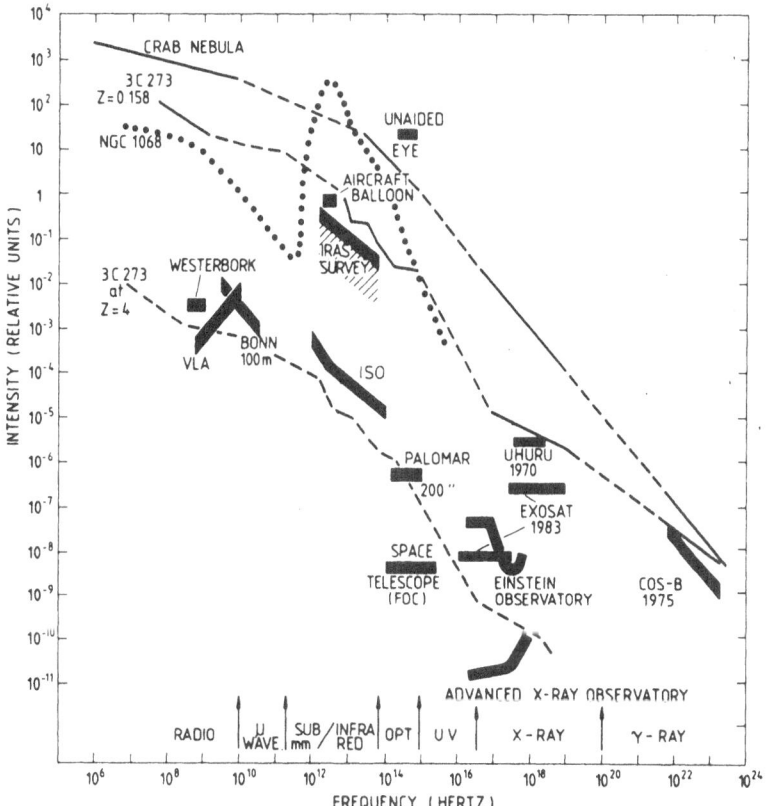

Figure 1 : Comparison made at ESA between the limiting sensitivity as a function of frequency of various astronomical facilities. Note that SIRTF is not included among the facilities.

Year	Radio	Infrared	Optical	UV	X-rays	γ-rays	In situ
2000	LDR ? **		SOLAR PROBE ? *,** / TRIO ? *				SOLAR PROBE ? *,**
1999		FIRST *	SAMSI ? **	UVSAT ? ***	LXAO ? ***		
1998					XMM *		
1997		SIRTF ? **			AXAF ? **		
1996					XAO ? ***		
1995	QUASAT *			LYMAN *		GRASP *	
1994			SOHO *,**	SOHO *,**	SOHO *,** / XTE **		SOHO *,**
1993	VSOP ? ***	ISO *	HRSO ? **	HRSO **	SXO ***		
1992				German 1m Telescope	SAX (I)		WIND **
1991	RADIOASTRON****			UARS ** / EUVE **	HESP ***		
1990							ULYSSES *,**
1989		COBE **	HIPPARCOS *		ROSAT (FRG)	GRO **	
1988			HST *,** / PHOBOS ****	IUE *,**	GRANAT **** / KVUANT ****	SIGMA ****	
1987					GINGA ***		

Table 1

*   ESA
**   NASA
***   Japan
****   Soviet Union

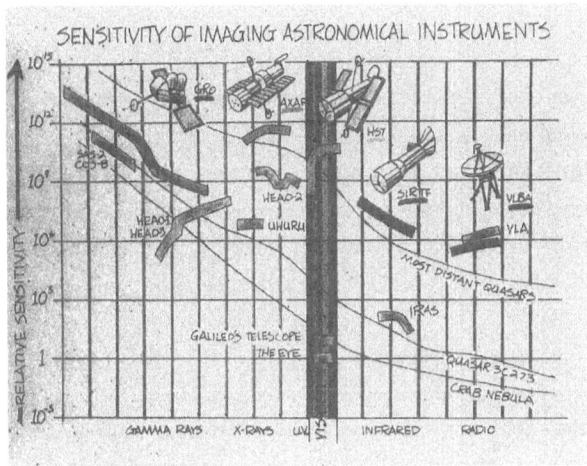

Figure 2 : Comparison similar to that of Figure 1 made at NASA. Note that ISO is not included among the facilities.

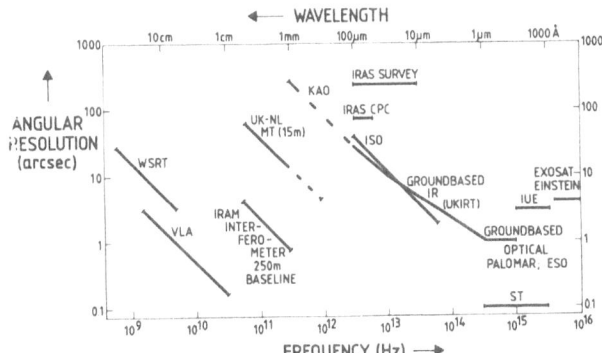

Figure 3 : Comparison between the angular resolution, as a function of frequency of various astronomical facilities.

We will now review briefly a few scientific areas where Space observations will yield major and fundamental progress.

III   Astrometry

Our present knowledge of absolute distances and absolute luminosities is based on the measurement of the parallaxes of a few nearby or very bright stars. However, this represents only a small sample on the Hertzprung-Russel diagram. The very brightest of the giants and main sequence stars are rare because the volume of space in wich a sufficient number of them can be found for any statistical study, is beyond the reach of ground based astrometry which is limited in accuracy, due to the detrimental effect of the Earth's atmosphere on the images.

One project, Hipparcos, programmed by ESA, will yield substantial progress and with an absolute accuracy of 2.10^{-3} arcsecond, will improve by nearly one order of magnitude the accuracy of ground based measurements. Hence, Hipparcos will bring within range of direct trigonometric parallaxes, many giant stars and stars which are located in the brighter end of the main sequence and which do not pertain to the solar neighbourhood. For the first time, direct luminosity calibration will be possible for regions of the Hertzprung-Russel diagram (Figure 4), and this will provide direct observational evidence for testing theories of stellar evolution.

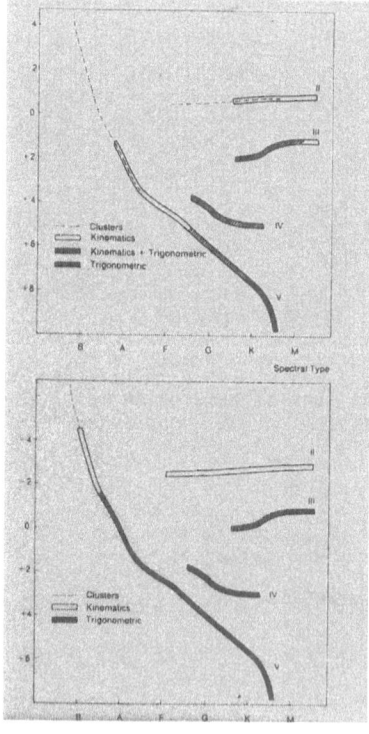

Figure 4 :

The absolute calibration of the Hertzprung-Russel diagram using the trigonometric parallaxes which will be obtained from Hipparcos (bottom).

The upper panel shows the present situation and evidences the progress gained with the help of Hipparcos.

The Hubble Space Telescope which is presently awaiting the resumption of shuttle launches, has a similar pointing accuracy as Hipparcos and its stellar sensors will in principle, be used for astrometry. There is a difference however, between the two missions. Hipparcos will provide an absolute reference frame in space to which the position of stars will be referred to, while the Hubble Space Telescope will provide measurements referred to a local reference frame only. Several observing programmes on the HST will however take advantage of this possibility and some are coupled with Hipparcos observations.

IV   Investigation of stellar interiors

The only direct information we have today on the physics of the interior of a star is from neutrinos originating in the core of the Sun where hydrogen is transformed into helium. It has been found that these solar neutrinos are under-abundant by a significant factor of three with respect to what is expected from the standard solar model, an observation which casts doubt on the model itself. Neutrinos from other stars are impossible to detect due to their very small flux, hence we should rely on other means of investigation.

Such means rest on the use of indirect information which is model dependent and allows to refine this model. The most promising one is helio or stellar seismology which diagnoses the oscillations of the main body of the Sun or of a star. Modes of oscillations can be detected in the power spectrum of the variations in luminosity of the tiny, periodic displacements of the Sun's surface. These oscillations are dependent upon the chemical composition, temperature and internal rotation of the Sun, and can be measured with very high resolution spectrometers. Such measurements have been made from the ground and from space with the ACRIM instrument on board the Solar Maximum mission and have yielded information, on the depth of the convective zone and of the rate of rotation of the interior of the Sun. More progress rest on the use of space techniques because of the detrimental effect of the Earth's atmosphere on the tiny doppler displacements or on the minute fluctuations in luminosity. In addition, a space mission if properly designed, can allow long series of uninterrupted observations increasing thereby the signal-over-noise ratio of the measurements.

In 1988, a consortium of experimenters under the leadership of Dr. C. Fröhlich in Switzerland, will fly a set of photometers on board the soviet mission PHOBOS which, on its journey to Mars and its moon, PHOBOS will benefit of

several months of full sunlight conditions. This will be the first time that high
accuracy uninterrupted measurements of solar luminosity oscillations will be made
from space. It is expected that the fundamental modes of the global oscillations
of the Sun will be fully revealed with this experiment.

The first cornerstone of ESA's Long Term Plan "Space Science-Horizon 2000"
is made of two projects : Cluster and SOHO. The latter is a Solar Observatory and
one of its prime goal is to study the interior of the Sun through helioseismology. Three types of instruments are foreseen :

- a high resolution velocity spectrometer, for the detection of global displacements of the Sun's surface ;

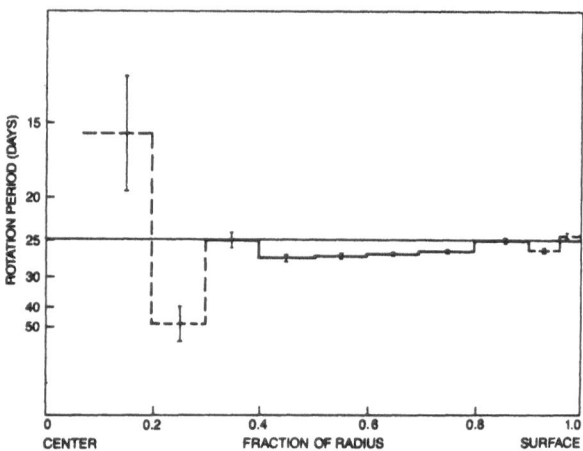

Figure 5 : Internal rotation of the Sun as a function of distance to the centre,
deduced from the differences in frequency seen for identical oscillation modes
propagating in opposite direction along the same meridian. The closer we go to the
centre, the larger the uncertainty of the "measurement". Going to space will allow
long period, low amplitude gravity waves, which come from the core, to be detected
as a consequence of elimination of atmospheric perturbations, yielding a much more
precise picture of the situation near Sun centre.

- a solar velocity imager for measuring the velocity field on the Sun's surface with a resolution of a few arcseconds ;

- a series of photometers and radiometers to measure the solar intensity fluctuations.

SOHO is to be launched in 1994, as part of a joint agreement with NASA.

In discussion is a similar project for stellar seismology, the so-called PRISMA project which could be mounted first on the future Space Station and then later on, on board a specially designed platform.

The successor of late NASA's solar optical telescope, the now called High Resolution Solar Observatory (HRSO), a 1m class telescope, will study the Sun with a resolution of 0.13 arcsecond, equivalent to one hundred km on the Sun's surface (Figure 6). Magnetic field measurements on board the HRSO of 10 gauss per resolution element, will permit a detailed analysis of the convective motions and their interaction with solar rotation, thereby providing a comprehensive study of the convection zone. HRSO is an international collaboration between the United States and Germany. Table 2 gives the main characteristics of the focal plane instruments.

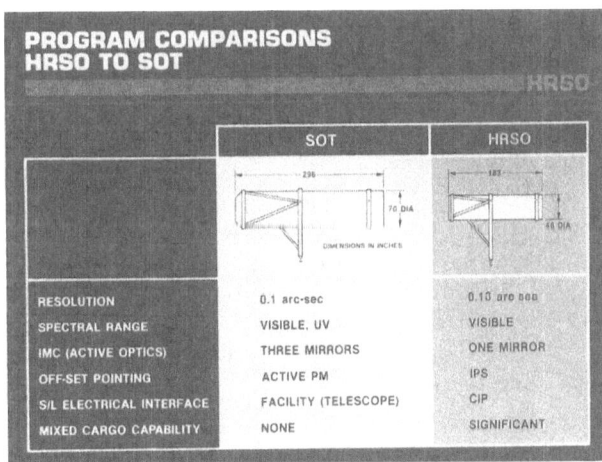

Figure 6 :

Characteristics of the High Resolution Solar Observatory as compared to those for SOT.

Table 2 : HRSO Follow-on

SPECTROGRAPH

Wavelength Intervals	2700–10000 A
Resolution	4 mA(@2800A)–16 mA(@10000A)
Field of View	0.10 X 76 arc sec
Exposures Times	0.1 -> ∞ sec
Polarization Analyzer	RCP, LCP, 3 linear $\begin{cases} \text{Linear} - 0 \\ \text{Linear} - 45 \\ \text{Linear} - 90 \end{cases}$

IMAGE MOTION COMPENSATION

Correction Range	+/- 15 arc sec
Bandwith	65 HZ
Stabilization	0.0188 arc sec RMS
Correl Track Error	0.005 arc sec RMS
Drift wrt Solar Rot	0.2 to 1.0 arc sec/hr

Table 2 : HRSO Focal Plane Instruments

PHOTOMETRIC FILTERGRAPH

Wavelength Interval	2200–8000 A
FWHM	0.4–200 A (fixed filters)
Accuracy	10% of FWHM
Field of View	152 X 188 arc sec
Exposure Times	0.001–10 sec
Film Capacity	25,600 frames (2.5 mil film)

TUNABLE FILTERGRAPH

Wavelength Interval	4600–7600 A
FWHM	40 mA(@4600A)–125 mA(@7600A)
Accuracy	1 mA relative; 5 mA absolute
Fields of View	152 X 152 arc sec (0.15" pixels)
	76 X 76 arc sec (0.075" pixels)
Typical Exposure	1 sec
Polarization Analyzer	RCP, LCP, 3 linear ⎰ Linear – 0 / Linear – 45 / Linear – 90 ⎱

At the Horizon of the next century, both ESA and NASA are planning to study the Sun with a solar probe or solar plunger which will fly in close vicinity (4 solar radii) to the solar surface (Figure 7). Tiny modifications in the probe's orbit may be induced by the Sun's gravity field. Hence, it will be possible to study the possible quadrupolar configuration of this field and in particular, the internal rotation through another technique than seismology. Such a measurement is also of fundamental importance as an experimental test of the theory of General Relativity.

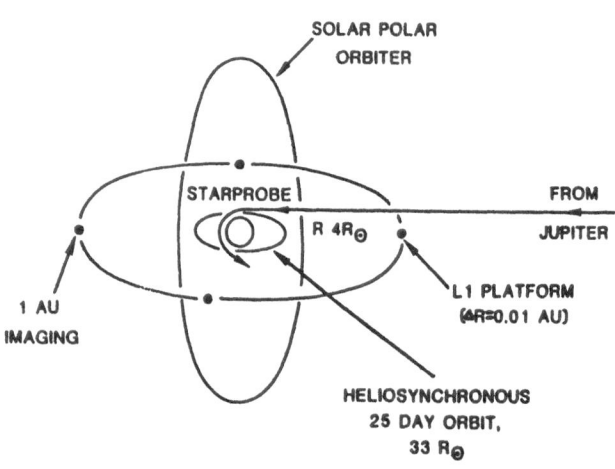

Figure 7 :

This drawing presents the missions that both ESA and the US National Academy of Sciences may plan in the future for solar observation. The existence of a solar probe (star probe on the drawing) travelling through the Corona at a distance of 4 solar radii is strongly advertised by both the Europeans and the Americans.

V   Investigation of external atmospheres and winds

Nearly 25 years of UV and X-rays observations of the Sun have allowed a refined study of its chromosphere and Corona. A phenomenological description is now available which evidences the prime role of motions and magnetic fields in governing the physics of these layers. However, a basic question remains today to be answered and this is the origin of the mechanisms which are at the origin of the heating of the Corona and how they depend on parameters, such as the radiation field, mass flows, convection and the magnectic field. It is likely that more progress will come from a comparison of the solar case, in which detailed resolved measurements are possible, with the case of stars for which a large variety of different conditions, like radiation field, rotatin rate, gravity, age etc., allows to study the influence of each of them on the heating rate. This interplay between solar and stellar physics is a key to further progress in this area.

UV and X-ray observations are only possible from Space. It is likely that SOHO will embark a set of UV spectrometers and coronographs (they are included in the model payload), with high resolution capabilities. However, it is the Space Station which, due to its orbit, will benefit mostly to solar observations (60 minutes of continuous observations per orbit). It may be equipped with high resolution spectrometers mounted at the focus of telescopes in the 40-60 cm telescope. An imaging Fourier spectrometer is foreseen on the station which will allow simultaneous two dimensional imaging in several very narrow spectral bands in the UV, visible and IR (Table 3).

As far as stars are concerned, Table 1 shows that the UV and X-rays spectrum will be studied by a wealth of instruments and we won't attempt to describe them all in detail. We will mention only Lyman, Rosat and XMM.

Lyman

Lyman is supposed to be a successor of the IUE which is still in operation after ten years of an uninterrupted and very successful mission and will probably remain operating until its on-board systems stop working. The main goal of Lyman will be the study of stellar chromospheres, winds and mass loss in stars of different galaxies through high spatial resolution (R = 30000) in the 90-120 mm spectral range where one finds the higher Lyman lines. The limiting magnitude will be V = 14. Contrary to the IUE, the orbit of the Lyman spacecraft won't be geostationary, but rather a high excentricity 48 hour 1000 x 120000 km orbit, which is optimal for the long observing times required to make up for the lower wavelength throughput. More details can be found in ESA - SCI.

Rosat

Rosat is a German satellite (Trumper, 1986) with substantial American (launch) and British contributions. It will carry a large X-ray telescope and a

Table 3

EXPERIMENTS PROPOSED FOR THE SPACE STATION (CORE AND COORBITING PLATFORM)

TITLE OF EXPERIMENT	MAIN FEATURES OF THE PROPOSED INSTRUMENT	PAYLOAD WEIGHT KG
IMAGING FOURIER TRANSFORM SPECTROMETER (IFTS)	• MOUNTED BEHIND A HIGH RESOLUTION TELESCOPE (DIAMETER 40 – 60 CM) • SIMULTANEOUS TWO-DIMENSIONAL FIELD OF VIEW WITH HIGH SPECTRAL RESOLUTION • SIMULTANEOUS IMAGING IN UV, V AND IR TO DETECT SHORT-TIME VARIATIONS	300
SOLAR SPECTROMETERS	FAR EUV SPECTROMETER	300
WHOLE SUN IMAGER	IMAGING IN SELECTED WAVELENGTH REGIONS	60
SOLAR IRRADIANCE MONITOR (SIM)	• IMMEDIATE IN-ORBIT COMPARISON OF SOLAR EUV IRRADIANCE WITH STANDARD SOURCE • SIMILAR IRRADIANCE COMPARISONS COVERING THE RANGE BETWEEN X-RAYS AND 400 MM	TBD
SUB MILLIMETRE WAVE SPECTROMETER (SMWS)	• EXTENSION OF MLS TECHNIQUE TO SUB MM WAVELENGTH RANGE TO MEASURE ADDITIONAL ATMOSPHERIC CONSTI-TUENTS	TBD
HELIUM-COOLED INFRARED SPECTROMETER (HCIRS)	• MEASUREMENT OF TRACE CONSTITUENTS OF THE MIDDLE AND UPPER ATMOSPHERE FROM THERMAL IR EMISSION • SPECTROMETER WITH A 3-STAGE COOLER, WAVELENGTH RANGE 4 – 150 MICRONS	300

X-UV telescope. The major objective is to perform the first all-sky survey in the X-ray (0.6-8 mm) and XUV (7 mm) bands by means of imaging telescopes. Other objectives will be to make detailed spectra of selected sources and of their variability. Table 4 gives the main characteristics of the instrumentation.

Figure 8 shows the sensitivity of the position sensitive proportional counter for the direction of point sources as a function of observing time as compared with that of Einstein and AXAF.

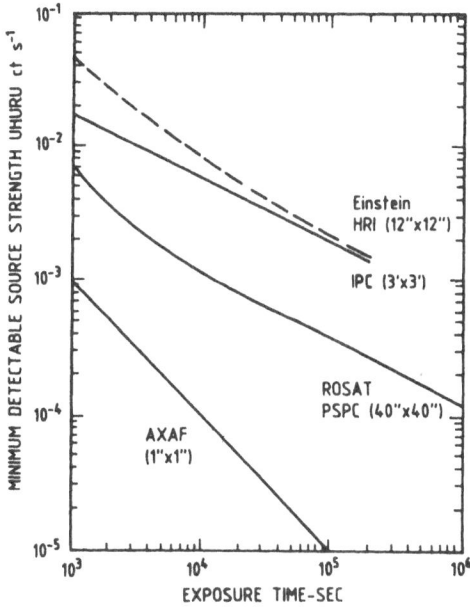

Figure 8 :

The sensitivity of the ROSAT position sensitive proportional counter for the detection of point sources as a function of observing time, as compared to the Einstein and AXAF capabilities.

Rosat is presently foreseen to be launched in 1989 by an american Delta Rocket.

XMM

The X-ray multi-mirror mission is the second cornerstone of ESA's Long Term Plan. The main characteristic of the instrumentation are given in Table 5 and Figure 9, compares the effective areas of 1 XMM mirror with that of the EXOSAT, Einstein and AXAF telescopes.

TABLE 4

ROSAT TELESCOPE INSTRUMENTATION

TELESCOPE WITH PSPC

FIELD OF VIEW	CIRCULAR 2° DIAMETER
EFFECTIVE COLLECTING AREA AT 1 KEV	420 CM^2
AT 0.28 KEV	470 CM^2
ON AXIS ANGULAR RESOLUTION AT 1 KEV	30 ARCSEC (FWHM)
0.28 KEV	1 ARCSEC (FWHM)

TELESCOPE WITH HRI

FIELD OF VIEW	36 ARCMIN
ON AXIS ANGULAR RESOLUTION	7 ARCSEC (HALF ENERGY WIDTH)

XUV WIDE FIELD CAMERA

FIELD OF VIEW	CIRCULAR 5° DIAMETER
ENERGY RANGE	0.21 - 0.041 KEV (60 - 300 Å)
RESOLUTION	2 ARCMIN RMS RADIUS (MEAN OVER FOV), 1 ARCMIN ON AXIS
TELEMETRY RATE	200 CTS/SEC.
ASPECT SOLUTION AND STABILITY	1 ARCMIN
FILTER BAND PASSES	LEXAN 60 - 170 Å)
	BE/LEXAN 112 - 190 Å) SURVEY AND POINTINGS
	AL/LEXAN 170 - 300 Å POINTINGS ONLY

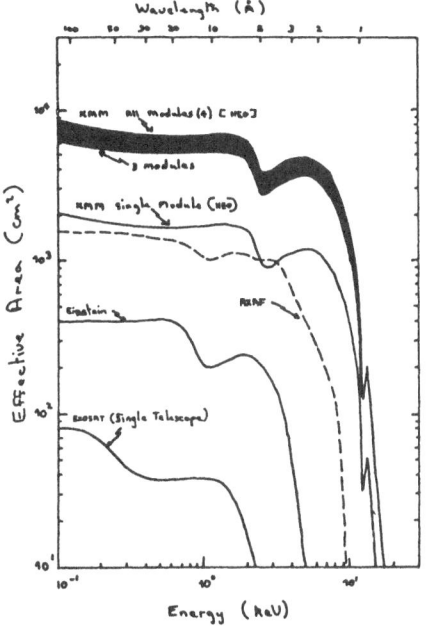

Figure 9 :

The effective area of a single and of four XMM mirror modules as a function of X-ray photon energy as compared with that of the EXOSAT, Einstein and AXAF telescopes.

One major part of the XMM observing programme will be devoted to the observation of stellar Coronae and in particular to the determination of their structure (both in their temperature and density), of their time variability (from minutes to days and years). It is hoped that a major breakthrough could be made on the problem of heating and how it depends upon the characteristic parameters of the star : rotation, age and total radiation field. For this, a high spectral resolution resolution is needed which requires the ghigh throughput of a mission like XMM. It will therefore be possible to resolve the temperature structure of the Corona through the use of pairs of lines which are density sensitive and of the differential emission measure analysis. The high throughput of XMM will allow the observation of objects located at distances at least 10 times larger than those where EXOSAT stars were observed.

As shown by EXOSAT, the continuous monitoring made possible by the use of a high excentricity orbit offers a considerable advantage. This is why a 60°, 1000 x 100000 km orbit has been selected. In that sense, XMM and AXAF are very complementary.

The launch of XMM is scheduled for 1998 on board an Ariane rocket. Direct in situ solar wind measurements will be made possible on board the SOHO and the NASA WIND satellite while Ulysses will be operating at high heliocentric latitudes above the ecliptic plane.

Table 5

OVERALL CHARACTERISTICS OF XMM

- Spatial resolution 30 arc sec HEW

- Effective area 10000 cm^2 at 2 keV ; 5000 cm^2 at 8 keV

- Broad band 0.2 - 10 keV spectroscopy with a resolving power of between 5 and 60

- Medium resolution spectroscopy between 0.1 - 3 keV with a resolving power of over 250

- High resolution spectroscopy in selected wavebands with a resolving power of over 1000

VI   Solar and stellar flares

Flares are amongst the strongest manifestations of solar activity and they find their origin in the release of magnetic energy stored in the Solar Corona. It is therefore in the X-ray, gamma-ray and radio regime that they can best be studied. The Solar Maximum mission and the Hinotori mission with their comprehensive set of instruments have indeed been the main source of progress in the understanding of this complex mechanism. The next cycle will be studied by a new japanese satellite, the "High Energy Solar Physics" mission or HESP (Watanabe, 1986). The HESP mission will be designed so as to improve on the SMM and Hinotori performance and particular emphasis will be given to high spatial resolution in the X-ray domain.

The key instruments on HESP will be the Hard X-ray Telescope (HXT) and the Soft X-ray Telescope (SXT) which are to resolve structures less than a few arc seconds in the Solar Corona. HXT will be a Fourier Synthetic Telescope while SXT which is built in the framework of a US-Japan cooperation is a grazing incidence mirror telescope, with multibroad band filters, able to provide images with a resolution in the 2-3 arcsecond range. Observations will be continuous and will allow the detection of the start of a Flare with an accuracy in absolute position better than the spatial resolution. This instrument should also permit to nearly image the whole sun. In addition, HESP will include a group of spectrometers covering from Soft X-rays to gamma-rays. Table 6 describes the main characteristics of HESP.

The SOHO mission of ESA will likely embark X-ray monitors, spectrometers and imagers which will observe solar flares. However, SOHO is not a flare dedicated mission and may only serve as a complement to a mission like HESP.

As far as stars are concerned, there are no specially dedicated missions, however most X-ray instruments on telescopes can be used for the study of flares and transients ; they are listed in column 5 of Table 1.

VII   Observations of stars at various phases of their evolution

We are dealing here with a statistical approach involving a large number of stars and obviously not one single object.

TABLE 6

HESP FOCAL PLANE INSTRUMENTATION

HARD X-RAY TELESCOPE (JAPAN, FOURIER SYNTHESIS TELESCOPE)

IMAGING CAPABILITY ABOVE 40 KEV AND IN SELECTED BANDS, BETWEEN 10 AND 100 KEV

SPATIAL RESOLUTION 8 ARCSEC > FWHM

TEMPORAL RESOLUTION 1 S.

POSITION SENSITIVE PHOTOMULTIPLIERS

SOFT X-RAY TELESCOPE (US-JAPAN, GRAZING INCIDENCE MIRROR TELESCOPE)

IMAGING CAPABILITY BETWEEN 10 - 100 Å

SPATIAL RESOLUTION 2 - 3 ARCSEC

TEMPORAL RESOLUTION BETTER THAN 1 S.

SPECTROMETERS FROM SOFT X-RAYS TO GAMMA RAYS

PHOTOMETRIC INTENSITY MONITOR

It happens that the medium and far infrared (including submillimetric wave-lengths) together with the X-UV and, gamma-ray domains carry information on both the early and late phases of the evolution of a star. The infrared opens the whole range of the interstellar cloud contraction and pre-main sequence phases, while the high energies give insight into the very late and neutron star state. Column 2 of Table 1 shows that in the infrared, a fairly large number of missions are envisaged in ESA and NASA. ESA is probably the most ahead with its ISO mission already in the development phase.

ISO, a 60 cm cryogenically cooled telescope (Figure 10) with a set of 4 focal plane instruments described in Table 7 will be a successor to the highly successful IRAS mission but contrary to the latter, will concentrate on observing a few preselected objects. ISO will play a central role in enlarging our under-standing of the problems associated with the formation of stars from moleculars clouds and dust cocoons. Spectroscopic instruments will allow the very strong winds associated with the young stars to be detected and investigated in detail.

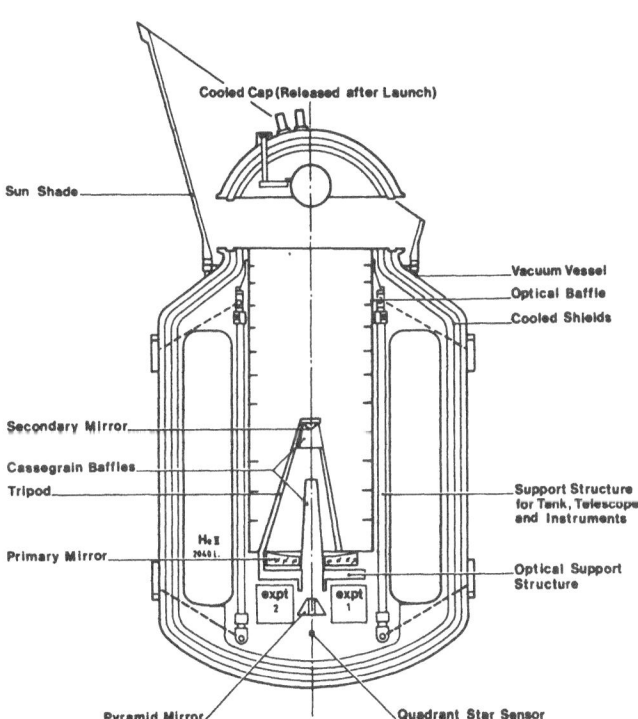

Figure 10 :

The Infrared Space Observatory (ISO) showing the telescope (60 cm) the cryostat and the main elements.

ISO Payload Module

The observation and spectroscopy of proto-planetary nebulae will provide a key to the still poorly known origin of our own planets. ISO will also permit, through molecular line spectroscopy to derive the physical properties of the atmospheres and circumstellar envelopes of cool stars of temperatures as low as about 2000 K. ISO will also provide means of observing the large circumstellar clouds which are formed around giant stars of late spectral types through mass loss since a substantial fraction of the total luminosity of these stars is usually converted into infrared continuum radiation by dust particles in these circumstellar shells.

Table 7

ISO Focal Plane Instrumentation

Instrument	Description	Wavelength Range (μm)	Principal Investigator
Camera (ISOCAM)	2-dimensional arrays for broad-& narrow-band imaging. Also polarimetry	3 - 17	C. Cesarsky Saclay, F
Short wavelength spectrometer (SWS)	Echelle grating with cross-dispersers and Fabry-Pérots for high resolution (R from $3000 - 10^5$)	3 - 50	Th. de Graauw Groningen, NL
Photo-polarimeter (ISOPHOT)	Multi-aperture, multiband photo-polarimetry. Also mapping & spectrophotometers at the shorter wavelengths	3 - 200	D. Lemke Heidelberg, D
Long Wavelength spectrometer (LWS)	Fabry-Pérot and grating combination, with resolution modes of $\simeq 200$ and $\simeq 10^4$	45 - 180	P. Clegg, QMC, GB

Later in the 1990's, ESA may launch one of the cornerstones of its long term plan, the submillimetric heterodyne spectroscopy mission also called FIRST, a 10 m telescope operating between 0.05 and a few mm. This instrument will be a unique tool to study protostars and the physics of the very early stage of star formation in particular, the link between molecular clouds and the already formed proto-stars, and the violent mass flows from early type stars at the early stages of their evolution. It will also allow to study the advanced stages of star formation where processed material is flowing into the interstellar medium. FIRST could be a precursor mission to the more ambitious Large Deployable Reflection (LDR) of 25 m envisaged later by NASA.

In the X-rays and gamma-rays, there is also a very large number of missions either approved or planned which will concentrate on neutron star and quasi perio-dic objects observations (see Columns 5 and 6 of Table 1).

In the framework of a cooperation between France and Soviet Union, the SIGMA project, a gamma-ray Telescope operationg between 30 KeV and 2 MeV and which will be launched in 1988 by a soviet launcher will concentrate its observations on the study of very compact galactic objects which represent the residu of massive stars at the ultimate phase of their evolution. SIGMA will also observe pulsars and X-ray binaries. The instrumentation is based on the principle of imagery through a coded mask technique and a set of 61 photomultipliers providing an angular resolu-tion of 1 to 13 arcmin., depending upon the source brightness. The operation modes will also allow spectroscopy over 1024 channels and variability studies. SIGMA will be placed on a high excentricity orbit 2000 x 200000 km which corresponds to a period of 4 days out of which 3.5 are free of radiation belt perturbations.

We should not of course forget that the main sequence is accessible through observation in the visible and the ultraviolet. Although it is likely that the Hubble Space Telescope will devote most of its time to extragalactic observations, faint stars will be observed. At ESA, Lyman, will mostly concentrate its observa-tions to stars (see also section V).

VIII Very high Resolution Imagery

It is well known that on the ground, it is not the quality of the optics which limit our ability to obtain better astronomical images but rather the

Earth's atmosphere which is highly turbulent and makes it very diffult to achieve resolution better than one arcsecond on the average and of a few tenths of an arcsecond in the best cases. Special techniques have been developed to overcome that stringent limitation such as interferometry and, more recently, speckle interferometry. These techniques as efficient as they are today, cannot compete however, as far as the search for ultra-high resolution in the micro-arcsecond range is concerned, with the so-called Very Long Baseline Interferometry technique. The latter is today limited by the length of the base which, for ground based telescopes cannot be larger than the diameter of the Earth. Hence, the strong advantage of space techniques which are able to overcome these limitations. Orbiting telescopes can operate over unlimited baselines and are unaffected by atmospheric turbulence. Their resolution is limited only by the intrinsic qualities of their optics and of the pointing systems which are used to maintain any astromical object in their field of view.

The first to take advantage of Space Techniques in VLBI radioastronomy will probably be the Soviets with their RADIOASTRON project which they plan to fly in the 1991-1994 timeframe. This will be a 10 m antenna orbiting at a maximum distance of 77000 km from the Earth which will "interfere" with ground based radio telescopes, providing a resolution in one direction of 30 micro arcseconds. The receivers will operate in the range of 0.33 and 22 GHz. A recent test (using telescopes) in the United States, Japan and Australia using the TDRSS (a data relay satellite) has proven the validity of the technique : out of 25 sources which were observed, 23 have provided interference fringes (Levy et al 1986). ESA is presently studying at phase A level a mission named QUASAT to be launched in the mid-1990's which is a 10 m offest antenna operating between 0.33 and 22 GHz, providing a baseline of 27000 km, yielding a resolution of 65 micro-arcseconds.

The Japanese are also envisaging a mission called VSOP which consists of a 5 m dish which will observe in the 5 and 22 GHz frequency bands from a low orbit with an inclination of 31 °. If approved, this mission will be launched in the 1993-1994 timeframe, to complement RADIOASTRON and possibly QUASAT later.

Extending the technique of radio VLBI to the optical or the near optical range and to shorter wavelengths is not easy an extrapolation. Such an attempt would be justified only as long as the technique would have reached on the ground, its own limitations and that it has been understood where these limitations come from. It is anticipated however that for baselines larger than a few tens of metres, the terrestrial atmosphere again will prevent maintaining phase coherence and that space techniques will be necessary to improve the resolution.

Two projects are being studied in Europe and in the United States : the TRIO (Figure 11) and SAMSI concepts. Both concepts, very similar in essence, are based on the use of two interfering telescopes and of a central station. The telescopes are 1 m in diameter separated by 10 km, which corresponds to a resolving power of 10 micro arcseconds, which is two orders of magnitude better than the capabilities of today's conventional interferometry. With this resolving power, the disk of a white dwarf would be resolved at a distance of 130 light years, and actual imaging of stellar surface is possible : in principle, a star of large angular diameter like Betelgeuse could be "seen" as the Sun with a seeing of one arcsecond.

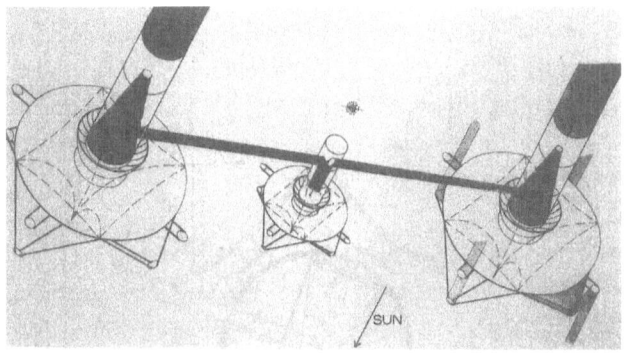

Figure 11 :

The TRIO concept studied at ESA including the two telescopes and the Central station. The three spacecrafts would be located at a lunar-solar Lagrangian point, where differential gravitational forces are weak.

The TRIO and SAMSI concepts differ essentially on the technique of propulsion to adjust the position of the three elements of the interferometer : chemical or ion propulsion for SAMSI, and solar radiation pressure for TRIO. The emphasis of the US project is on sweeping across as many sources as possible in the minimum time while the Europeans have concentrated on obtaining two dimensional images.

In a recent study undertaken for NASA, the National Academy of Science has identified Space interferometry as a major element of Space astronomy in the 21st century and two projects in particular look very promising although at the limit of today's human ambition :

- a large space telescope array made of 8 mirrors of 8 m each providing a resolution of 25 milliarcseconds at 10 microns and,

- a long baseline optical interferometre made of two or more 8 m telescopes operating on a baseline of 100 km, reaching an angular resolution in the micro-arcsecond range.

As far as the Sun is concerned, its proximity to us, implies that Ultra High Resolution Imagery (UHRI) deals with structures of only a few km or even hundreds of meters at the solar surface. One may therefore question whether there are any such structures in the Sun and whether they can be observed since the photons mean free path is much longer than the characteristic dimension of these hypothetical structures. In the framework of the already mentionned NAS study, a workshop was organized in January 1986 in Tucson, which involved a substantial participation of European scientists, with the very precise goal of assessing the scientific value of UHRI for the Sun, and of analyzing ways and technical means which could be used to that effect. The proceedings of this workshop are available (High Spatial Resolution Observations of the Sun in the 21st century, D. Rabin, J. Leibacher Ed. Tucson NOAO, Jan. 1986). The outcome of the study was to confirm definitely the scientific interest of UHRI, especially because there may be some structures totally confined by the magnetic field into small volumes and whose radiative output might be strong enough to be detectable.

Several techniques have been proposed to obtain the images, such as operating at short wavelengths to beat down the diffraction problem of telescopes, by using :

- very high quality grazing incidence X-ray mirrors,
- or normal incidence optics coated with multi-reflecting layers, operating near diffraction limit in the far UV and X-ray domain,
- or phase plane correctors (such as in the DILUVIS concept of Damé, Foing et al (1986).

Speckle Imaging techniques and aperture synthesis have also been considered, as well as interferometry using several telescopes or a remake of Michelson's experiment using the late SOT or its successor HRSO. A small preparatory experiment, (Damé, Foing et al 1986) has even been proposed as a test using 4 x 10 cm telescopes over a baseline of 2 to 4 m, the so-called SOABI concept to be performed on board one of ESA's Eureca platforms or on the Space Station.

IX   Conclusion

As it can be easily judged from the previous enumeration of projects and of scientific objectives, the future prospects for stellar and solar observations are very bright, provided the delays induced by the present crisis in Space launchers will soon be brought to a stop. From the deep interior of stars to the most external parts of their atmospheres, Space techniques offer unique means of investigation. The major world space agencies have their own plans established in this area but it can easily be realised that none of them would be able, alone, to provide the complete and necessary weaponery which we have briefly described in this short article and which would yield to the most advanced progress in our understanding of stellar and solar physics.

However, by joining their efforts through close consultation and cooperation, they might altogether contribute to the most impressive endeavour in the history of astronomy. This peaceful enterprise can be put at work at any time and the plans in each agency are ambitious enough to prevent anyone of them to impose any kind of leadership in this area. The recent example of Halley's comet space missions which followed exactly that track is a clear indication that cooperation in the envisaged space missions would be the most promising approach to solve the exciting problems of stellar and solar physics.

References

L. Damé, B. Foing, J.-C. Vial, and Bourdet G., 1986, Proceedings of the BNSC/ESA/CNES Workshop on Solar and Terrestrial Physics on Space Stations / Columbus, Rutherford Appleton Laboratory, Chilton UK, 14-15 October 1986.

G.-S. Levy, R.-P. Linfield, J.-S. Ulvestad, C.-D. Edwards, J.-F. Jordan, Jr, S.-J. Di Nardo, C.-S. Christensen, R.-A. Preston, L.-J. Skjerve, L.-R. Stavert, B.-F. Burke, A.-R. Whitney, R.-J. Cappallo, A.-E.-E. Rogers, K.-B. Blaney, M.-J. Maher, C.-H. Ottenhoff, D.-L. Jauncey, W.-L. Peters, T. Nishimura, T. Hayashi, T. Takano, T. Yamada, H. Hirabayashi, M. Morimoto, M. Inoue, T. Shiomi, N. Kawaguchi, H. Kunimori, "Very Long Baseline Interferometric Observations made with an Orbiting Radio Telescope", 1986, Science, 234, page 187.

J. Trumper, 1986 "Cosmic radiation in contemporary Astrophysics", M.M. Shapiro (ed), Reidel Publishing company, 241-247.

T. Watanabe 1986, Proceedings of the 2nd Japan-China Workshop on Stellar activities and Observational Techniques, 1986, Kyoto.

POSTER CONTRIBUTIONS

Late-type stars: Dynamo, rotation etc.

Emerging trends in magnetic field measurements for late type stars:
Empirical constraints on stellar dynamo theories, S.H. SAAR, J.L. LINSKY

Nonlinear models of stellar dynamos, G. BELVEDERE, R.M. PITADELLA,
M.R.E. PROCTOR

Lithium depletion and rotation for solar type stars in the Hyades,
R. REBOLO, J.E. BECKMAN

The closest "solar analogs" in the "bright star catalogue", H. NECKEL

The Sun as a star, global properties of the Sun

A Belgrade program for monitoring of activity - sensitive spectral lines
of the Sun as a star, J. ARSENIJEVIC, M. KARABIN, A. KUBICELA, I. VINCE

Solar radiation and solar rotation, H. WÖHL

Asymmetry and variations of solar limb darkening along the diameter
defined by diurnal motion in April 1981, H. NECKEL, D. LABS

Periodicities of the sunspot areas during the solar cycle, J. PAP

Evidence for a north-south asymmetry in the rotation of the solar magnetic
fields during two solar cycles, E. ANTONUCCI, J.T. HOEKSEMA, P.H. SCHERRER

On the longitudinal distribution of the solar background magnetic fields,
V. BUMBA, L. HEJNA

Flux emergences as a consequence of local solar rotation anomalies,
Z. MOURADIAN

Observational constraints for solar dynamo, A. BRANDENBURG, I. TUOMINEN

A possible explanation of the global resonances in the solar magnetic field, P. HOYNG

Is the solar differential rotation a manifestation of the pure zonal flow?, P. AMBROZ

On low-modal stationary torsional waves in the solar differential rotation, L. HEJNA

A global network of solar seismology stations, Y.P. ELSWORTH, G.R. ISAAK, S.M. JEFFERIES, C.P. McLEOD, R. NEW, H.B. VAN DER RAAY

The effect of a localized magnetic field on p-mode frequencies, W.R. CAMPBELL, B. ROBERTS

Small-scale velocity fields, granulation

Short-term profile variations of photospheric lines, M.T. GOMEZ, C. MARMOLINO, G. ROBERTI, G. SEVERINO

Centre-to-limb variation of granular velocity fields, R. KOMM, W. MATTIG

Variation of the granulation with the solar latitude, M. COLLADOS, I. RODRIGUEZ HIDALGO, M. VAZQUEZ

Limb effect in neutral iron lines, B.N. ANDERSEN

Are solar granules convective features?, R. MULLER, T.H. ROUDIER, J.M. MALHERBE, P. MEIN

Granular convection in the Sun and Procyon, M. STEFFEN

Radiative convection in the Eddington approximation, J.M. EDWARDS

Sunspots, active regions

Fragments in sunspots, J.I. GARCIA DE LA ROSA

Oscillations in sunspots, H. BALTHASAR, G. KÜVELER, E. WIEHR

Motions around a decaying sunspot, R. MULLER, B. MENA

The dynamics of active regions in the solar atmosphere, K.H. FLEIG,
W. MATTIG, A. NESIS

Small-scale magnetic fields, flux tubes

On the interaction between magnetic fields and granulation, J.I. GARCIA DE
LA ROSA, M. COLLADOS

Continuum intensity and magnetic flux of solar fluxtubes, J.C. DEL TORO
INIESTA, M. SEMEL, M. COLLADOS, J. SANCHEZ ALMEIDA

Heights of formation of Stokes profiles in solar magnetic fluxtubes,
B. LARSSON, S. SOLANKI

Model atmospheres in intense flux tubes, S.S. HASAN

Oscillations in magnetix flux tubes, S.S. HASAN, Y. SOBOUTI

Surface and body waves in magnetic flux tubes, T. ABDELATIF

Dynamical evolution of thin magnetic flux tubes, K. ZÄHRINGER,
P. ULMSCHNEIDER

The interaction of whirlflows with concentrated magnetic fields,
V. ANTON

Adaptive mapping methods for compressible MHD, E. WEISSHAAR

Flux tube models with full radiative transport, U. GROSSMANN-DOERTH,
M. KNÖLKER, M. SCHÜSSLER, E. WEISSHAAR

Generation of longitudinal magnetohydrodynamic tube waves in stellar
convection zones, P. ULMSCHNEIDER, Z. MUSIELAK, R. ROSNER

SESSION III

Chromospheric structures, prominences

A two-dimensional model for a solar prominence, J.L. BALLESTER,
E.R. PRIEST

Observations of prominence-like clouds in the corona of a rapidly-
rotating G-K dwarf, A. COLLIER CAMERON, R.D. ROBINSON

The evolution of discrete structures in the atmosphere of AR Lacertae
between 1980 and 1985, J.E. NEFF, J.L. LINSKY, M. RODONO, F.M. WALTER

Chromospheres, transition regions, coronae:
Heating and emission

Calibration of the Ca II K line-core flux density and the magnetic flux
density in stars using solar data, C.J. SCHRIJVER, J. COTÉ

Looking at the past history of the Sun: A comparison with the young solar-
type star 53 Aquarii, M. CERRUTI-SOLA, R. PALLAVICINI, L. PASQUINI

Empirical relations between photospheric magnetic fluxes and atmospheric
radiative losses for cool dwarf stars, C.J. SCHRIJVER, S.H. SAAR

Ca II emission in old main sequence stars from ESO 1.5 m Coudé spectra,
E. MARILLI, S. CATALANO, C. TRIGILIO

Ca II emission in young open clusters from low resolution Reticon spectra,
C. TRIGILIO, S. CATALANO, E. MARILLI, V. REGLERO

Observation at chromospheric and temperature minimum level of periodic
meso-scale brightenings, M. MARTIC, L. DAMÉ

Elemental abundances in different solar regions from EUV observations, G. NOCI, D. SPADARO, R.A. ZAPPALA, F. ZUCCARELLO

XUV active coronal regions on the Sun, M. LANDINI, B.C. MONSIGNORI FOSSI

Coronal heating by evolution of a sunspot magnetic field, A.M. DIXON, P.K. BROWNING, E.R. PRIEST

Mode coupling of continuum modes in 2D coronal loops and arcades, S. POEDTS, M. GOOSSENS

Ultraviolet emission from the Sun and stars: A comparison of IUE and Skylab Spectra, A. CAPPELLI, M. CERRUTI-SOLA, C.C. CHENG, R. PALLAVICINI

Width-luminosity relations of EUV emission lines in late-type stars, O. ENGVOLD, Ø. ELGARØY

Full disk X-ray observations of the Sun with the Einstein Observatory, J.H.M.M. SCHMITT

X-ray and UV emission from late type stars. Coronal structure and energy balance, M. LANDINI, B.C. MONSIGNORI FOSSI, F. TRIBOLI, R. PALLAVICINI

About the coronal activity cycles of the Sun, S. KOUTCHMY, J.-C. NOENS

Rotation and short periodicities of the green corona for the Sun as a star, V. RUSIN, M. RYBANSKY, J. ZVERKO

Inner solar wind region

Coronal scintillations with water masers, M.K. BIRD, M. PÄTZOLD, R. GÜSTEN, W. SIEBER, N.A. LOTOVA

On the possibility of measuring the amplitude of Alfvén waves in the inner solar wind region, R. ESSER, E. LEER, T.E. HOLZER

Energetic events, flares, flare stars

Turbulence and velocity fields in H_α solar ejecta, P. MEIN, N. MEIN

Transition zone and corona signatures of chromospheric mass ejection, B. SCHMIEDER

Periodic behaviour in sudden disappearances of solar prominences, G. VIZOSO, J.L. BALLESTER

The necessity of active region filaments for two ribbon flares, A. HOOD, U. ANZER

Possible mechanisms for flare onset in line-tied coronal arcades, J.P. MELVILLE, A.W. HOOD, E.R. PRIEST

Resistive tearing modes in line-tied coronal magnetic fields, M. VELLI, A.W. HOOD

Effects of electron bombardment on the low atmosphere of the Sun during flares, J. ABOUDARHAM

Solar millisecond radio spikes in the decimetric range, A.O. BENZ, M. GÜDEL

How far does the analogy between the solar and stellar flares work?, A. GRANDPIERRE

Modelling of 2-ribbon flares on the Sun and stars: Application to flares on EQ Peg and Prox Cen, G. POLETTO, R. PALLAVICINI, R.A. KOPP

Exosat observations of flare stars: Implications for the heating of solar and stellar coronae?, R. PALLAVICINI

Radiation losses in chromospheric emission lines of solar neighbourhood flare stars, B.R. PETTERSEN

Some new results on the Pleiades flare stars, J. KELEMEN

F-stars

Activity in warm stars, F.M. WALTER, C.J. SCHRIJVER

RS Canum Venaticorum systems

ubvy and H_α, H_β Strömgren photometry of RSCVn systems,
M.J. ARÉVALO, C. LÁZARO, J.J. FUENSALIDA, V. REGLERO

Infrared evidence of dust around RSCVn systems, C. LÁZARO, M.J. AREVALO

Coronal activity of RSCVn systems, O. DEMIRCAN

Radio outbursts in HR 1099: Quantative analysis of flux spectrum and
intensity distribution, K.-L. KLEIN, F. CHIUDERI-DRAGO

Dividing lines, mass loss

Coronal and noncoronal stars, R. HAMMER

Hot loops, cool loops and the coronal dividing line, B.M. HAISCH,
A. MAGGIO, G.S. VAIANA

Blue-red asymmetries in uncontaminated MgII h and k emissions from
giants: A reliable outflow diagnostic, G. VLADILO, J.E. BECKMAN,
L. CRIVELLARI, B. FOING

Episodic mass loss in late-type stars due to acoustic wave packets,
M. CUNTZ

Pre-main sequence stars

Chromospheric variability mechanisms affecting pre-main sequence stars,
A. BROWN

Effects of stellar chromospheres on spectral classification,
I. APPENZELLER

Emission lines in southern pre-main sequence stars, A. MAGAZZU,
M. FRANCHINI, R. STALIO

SESSION IV

Ultraviolet solar irradiance measurement from 200 to 358 nm during
Spacelab 1 mission, D. LABS, H. NECKEL, P.C. SIMON, G. THUILLIER

Solar and Heliospheric Observatory, SOHO, V. DOMINGO

The CHASE experiment on Spacelab 2, J. LANG, E.R. BREEVELD,
B.E. PATCHETT, J.H. PARKINSON, A.H. GABRIEL, J.L. CULHANE,

Special Presentations

The China 35 cm Solar Telescope and its future, LI TING

PARTICIPANTS

ABDELATIF, T.	London, U.K.
ABOUDARHAM, J.	Meudon, France
AMBROZ, P.	Ondrejov, CSSR
ANDERSEN, B.N.	Noordwijk, Netherlands
ANTON, V.	Freiburg, F.R.G.
ANTONUCCI, E.	Turin, Italy
APPENZELLER, I.	Heidelberag, F.R.G.
ARSENIJEVIC, J.	Belgrad, Yougoslavia
BALTHASAR, H.	Göttingen, F.R.G.
BELVEDERE, G.	Catania, Italy
BENZ, A.O.	Zürich, Switzerland
BERTON, R.	Meudon, France
BLANCHOUD, F.	Bern, Switzerland
BONNET, R.	Paris, France
BRANDT, P.	Freiburg, F.R.G.
BRANDENBURG, A.	Helsinki, Finland
BROWN, A.	Boulder, Colorado
BRUGGMANN, G.	Bern, Switzerland
CAMERON, A.	Herstmonceux, U.K.
CATALANO, S.	Catania, Italy
CHAMBE, G.	Paris-Meudon, France
CHIUDERI-DRAGO, F.	Meudon, France
CHMIELEWSKI, Y.	Genf, Switzerland
COLLADOS, M.	La Laguna, Tenerife, Spain
COVINO, E.	Napoli, Italy
CUNTZ, M.	Heidelberg, F.R.G.
DAME, L.	Verrière-le-Buisson, France
DANIEL, H.U.	Heidelberg, F.R.G.
DEMIRCAN, O.	Ankara, Turkey
DEUBNER, F.-L.	Würzburg, F.R.G.
DIXON, A.M.	St. Andrews, U.K.
DOMINGO, V.	Noordwijk, Netherlands
DRAVINS, D.	Lund, Sweden
EDWARDS, J.M.	Cambridge, U.K.
ELGARØY, Ø.	Oslo, Norway
ESSER, R.	Tromsø, Norway
EVANS, D.	St. Andrews, U.K.

FERRIZ MAS, A.	Freiburg, F.R.G.
FLECK, B.	Würzburg, F.R.G.
FLEIG, K.H.	Freiburg, F.R.G.
FREIRE FERRERO, R.	Strasbourg, France
GABRIEL, A.H.	Verrières le Buisson, France
GOMEZ, M.T.	Napoli, Italy
GOOSSENS, M.	Heverlee, Belgium
GRAETER, M.	Bern, Switzerland
GRANDPIERRE, A.	Budapest, Hungary
GROSSMANN-DOERTH, U.	Freiburg, F.R.G.
GÜDEL, M.	Zürich, Switzerland
HAISCH, B.M.	Palo Alto, California, USA
HAMMER, R.	Freiburg, F.R.G.
HANG, E.	Tübingen
HANKUS, M.	Kraków, Poland
HASAN, S.S.	London, U.K.
HEJNA, L.	Ondrejov, CSR
HENOUX, J.C.	Paris-Meudon, France
HERMANS, D.	Heverlee, Belgium
HOOD, A.	St. Andrews, U.K.
HOYNG, P.	Utrecht, Netherlands
HUBER, M.C.E.	Zürich, Switzerland
JARDINE, M.	St. Andrews, U.K.
JASZCZEWSKA, M.	Kraków, Poland
KELEMEN, J.	Budapest, Hungary
KLEIN, K.-L.	Paris-Meudon, France
KNEER, F.	Göttingen, F.R.G.
KNÖLKER, M.	Göttingen, F.R.G.
KOMM, R.	Freiburg, F.R.G.
KOUTCHMY, S.	Sunspot, New Mexico
KUPERUS, M.	Utrecht, Netherlands
LABHARDT, L.	Basel, Switzerland
LABS, D.	Heidelberg, F.R.G.
LANDINI, M.	Firenze, Italy
LANG, J.	Chilton, U.K.
LARSSON, B.	Lund, Sweden
LAUFER, J.	Würzburg, F.R.G.

LEROY, B.	Meudon, France
LI TING	Freiburg, F.R.G.
LINSKY, J.L.	Boulder, USA
LOTHIAN McKAY, R.	St. Andrews, U.K
LUSTIG, G.	Graz, Austria
MAGUN, A.	Bern, Switzerland
MAGUN, Ch.	Bern, Switzerland
MALHERBE, J.M.	Meudon, France
MARCO-SOLER, E.	Freiburg, F.R.G.
MARTI, H.	Bern Switzerland
MARTIC, M.	Paris, France
MATTIG, W.	Freiburg, F.R.G.
McWHIRTER, R.W.M.	Chilton, U.K.
MEIN, N.	Meudon, France
MEIN, P.	Meudon, France
MELVILLE, J.P.	Edinburgh, U.K.
MERCIER, C.	Meudon, France
MEYER, J.-P.	Geneva, Switzerland
MONSIGNORI-FOSSI, B.	Firenze, Italy
MONTMERLE, T.	Gif sur Yvette, France
MOURADIAN, Z.	Meudon, France
MÜLLER, E.A.	Basel, Switzerland
MULLER, R.	Bagnères de Bigorre, France
NECKEL, H.	Hamburg, F.R.G.
NEFF, J.E.	Boulder, USA
NESIS, A.	Freiburg, F.R.G.
NEW, R.	Birmingham, U.K.
NOVOCKY, D.	Tatranskà Lomnica, CSSR
PÄTZOLD, M.	Bonn, F.R.C.
PAHLKE, K.-D.	Göttingen, F.R.G.
PALLAVICINI, R.	Firenze, Italy
PAP, J.	Budapest, Hungary
PATERNÓ, L.	Catania, Italy
PECKER, J.-C.	Paris, France
PETTAUER, J	Kanzelhöhe, Austria
PETTERSEN, B.R.	Oslo, Norway
PITADELLA, R.M.	Catania, Italy

POEDTS, S.	Heverlee, Belgium
POLETTO, G.	Firenze, Italy
PRIEST, E.R.	St. Andrews, U.K.
REIMERS, D.	Hamburg, F.R.G.
RENAN DE MADEIROS, J.	Sauverny-Versoix, Switzerland
RIBES, E.	Meudon, France
RIGHINI, A.	Firenze, Italy
ROBERTS, B.	St. Andrews, U.K.
ROBERTSON, J.A.	St. Andrews, U.K.
RODONO, M.	Catania, Italy
ROUDIER, Th.	Toulouse, France
ROXBURGH, I.W.	London, U.K.
RUTTEN, R.	Utrecht, Netherlands
SANCHEZ ALMEIDA, J.	La Laguna, Tenerife, Spain
SCHLEICHER, H.	Freiburg, F.R.G.
SCHMALZ, S.	Freiburg, F.R.G.
SCHMIEDER, B.	Meudon, France
SCHMIDT, W.	Freiburg, F.R.G.
SCHMITT, J.	Garching, F.R.G.
SCHOBER, H.-J.	Graz, Austria
SCHRIJVER, C.J.	Boulder, USA
SCHRÖTER, E.-H.	Freiburg, F.R.G.
SCHÜSSLER, M.	Freiburg, F.R.G.
SEVERINO, G.	Napoli, Italy
SKALEY, D.	Freiburg, F.R.G.
SKOPAL, A.	Tatranskà Lomnica, CSSR
SMALDONE, L.	Napoli, Italy
SOLANKI, S.	Zürich, Switzerland
SOLTAU, D.	Freiburg, F.R.G.
STÄHLI, M.	Bern, Switzerland
STEELE, C.D.C.	St. Andrews, U.K.
STEFFEN, M.	Kiel, F.R.G.
STEIGER, R.V.	Bern, Switzerland
STENFLO, J.O.	Zürich, Switzerland
STIX, M.	Freiburg, F.R.G.
TITLE, A.M.	Palo Alto, Ca, USA
DEL TORO INIESTA; J.C.	La Laguna, Tenerife, Spain

TOZZI, G.P.	Firenze, Italy
TRIBOLI, F.	Firenze, Italy
TRUJILLO BUENO, J.	Göttingen, F.R.G.
TUOMINEN, I.	Helsinki, Finland
UITENBROEK, H.	Utrecht, Netherlands
ULMSCHNEIDER, P.	Heidelberg, F.R.G.
VELLI, M.	St. Andrews, U.K.
WAGNER, K.-D.	Heidelberg, F.R.G.
WEISS, N.O.	Cambridge, U.K.
WEISSHAAR, E.	Bayreuth, F.R.G.
WHITE, S.	Garching, F.R.G.
WITTMANN, A.	Göttingen, F.R.G.
WÖHL, H.	Freiburg, F.R.G.
WÜLSER, J.-P.	Bern, Switzerland
LIN YUN	Lissbon, Portugal
ZÄHRINGER, K.	Heidelberg, F.R.G.
ZAHN, J.-P.	Toulouse, France
ZAYER, I.	Zürich, Switzerland
ZWAAN, C.	Utrecht, Netherlands

Lecture Notes in Mathematics

Lecture Notes in Physics

H. Scheffler, H. Elsässer

Physics of the Galaxy and Interstellar Matter

Translated by A. H. Armstrong

Astronomy and Astrophysics Library

1987. 207 figures. XI, 492 pages.
ISBN 3-540-17314-5

This book, based on the author's long-standing experience in teaching astronomy courses, presents the complete modern picture of the physics of the Milky Way system. The first part of the book deals in comprehensible terms with topics of more empirical character, such as the positions and motions of stars, the structure and kinematics of the stellar system and interstellar phenomena. The more advanced second part is devoted to the interpretation of observational results, i.e. to the physics of interstellar gas and dust, stellar dynamics, the theory of spiral structures and the dynamics of interstellar gas.

H. Karttunen, P. Kröger, H. Oja (Eds.)

Fundamental Astronomy

1987. 399 illustrations including 36 color plates.
XIII, 478 pages. ISBN 3-540-17264-5

This book is an ideal introductory text for first- and second-year university students and for serious amateurs in the field. But the text has been arranged so that it can be understood without the mathematics, so the book is also of interest to a wider audience.

Springer-Verlag
Berlin Heidelberg New York
London Paris Tokyo

S. Laustsen, C. Madsen, R. M. West

Exploring the Southern Sky

A Pictorial Atlas from the European Southern Observatory (ESO)

1987. 240 photographs, partly in colour,
31 diagrams and a fold-out plate. VI, 274 pages.
ISBN 3-540-17735-3
German edition available from Birkhäuser Verlag

Extensive captions by an expert author team, 30 reference maps, tables of plate data, and indexes complement the images and enable readers to locate and understand the significance of each astronomical object. Of special interest are photographs of Supernova 1987 A and Halley's Comet and a four-foot, panoramic poster of the Milky Way as it appears to a nighttime observer in the Southern Hemisphere.

S. M. Faber (Ed.)

Nearly Normal Galaxies

From the Planck Time to the Present

The Eighth Santa Cruz Summer Workshop in Astronomy and Astrophysics, July 21 – August 1, 1986, Lick Observatory.

1987. 133 figures, 7 tables. XXVI, 464 pages.
ISBN 3-540-96521-1

Contents. Stellar Evolution in Galaxies. – Small Objects. – Galactic Structure and Dynamics: Observations. – Galactic Structure and Dynamics: Theory. – Global Parameters of Galaxies. – Galaxies in Relation to Larger Structures. – Distant Galaxies. – Dark Matter. – Galaxies Before Recombination. – Galaxies After Recombination. – Conference Summary.

Springer